Weeds of Canada

CLARENCE FRANKTON
and
GERALD A. MULLIGAN

Illustrations by
W. H. WRIGHT
and
ILGVARS STEINS

Publication 948

Published by
NC Press Limited
in cooperation with
AGRICULTURE CANADA
and the
Canadian Government Publishing Centre
Supply and Services Canada

First Printing 1955
Revised 1970
Reprinted 1971, 1974, 1975, 1977, 1980
Revised 1987

© Minister of Supply and Services Canada — 1987

Catalogue number: A43-9/1986E

Canadian Cataloguing in Publication Data

Frankton, Clarence
  Weeds of Canada

Rev.
ISBN 1-55021-016-5

1. Weeds - Canada. 2. Weeds - Canada - Identification. I. Mulligan, Gerald A., 1928-. II. Canada. Agriculture Canada. III. Title.

SB613.F73 1987	581.6'52'0971	C87-094870-7

We would like to thank the Ontario Arts Council and the Canada Council for their assistance in the production of this book.

New Canada Publications, a division of NC Press Limited, Box 4010, Station A, Toronto, Ontario M5H 1H8.

Printed and bound in Canada

CONTENTS

Weeds of Canada

INTRODUCTION

Some people will use this manual because they realize that knowing the name of a weed is a key to information available on the control of that particular plant. Others will have an interest in weeds as part of their environment or as classroom studies. Whatever the purpose of the reader, this publication will assist him in identifying the important weeds found in Canada and will supply general information that will be of interest.

More than a thousand different plants, nearly a fifth of all the species found in Canada, can be designated as weedy. As it is impossible to introduce all these weeds in a single volume no such attempt has been made. Assistance in the selection of the more limited number included in this manual has been freely given by weed specialists and agriculturists across Canada.

A total of 230 species are described in the text. The 101 plates include complete or partial drawings of 154 species.

LOSSES DUE TO WEEDS

In Canada, a conservative estimate of the annual losses caused by weeds is about 200 million dollars. Possibly this amount should be much higher. In the United States, farmers' annual losses from weeds have been estimated at about 5 billion dollars. Based on the relative hectarages under crop in the United States and in Canada, and assuming that the ratio of loss is similar, the Canadian loss could be over 500 million dollars. There seems to be general agreement that the losses caused by weeds are greater than the combined losses produced by animal diseases, plant diseases, and insect pests.

Weeds take their toll on farms in many ways. Weeds compete directly with the growing crop for light, moisture, and minerals, and the reduction of crop yield because of this competition is one of the main sources of loss. Weed control is responsible for direct outlays in chemicals and machines and for a large portion of the man-hours spent in tilling and harvesting. The cost of handling weed seeds in cereals and the cost of transport, dockage, and cleaning are important expenses. Less direct losses result from weeds harboring insects or disease organisms, which attack crops. Heavy stands of noxious weeds depreciate land values. Weeds also cause losses or unthriftiness of livestock when poisonous ones are eaten, and reduce the value of livestock products, such as wool and hides contaminated by burs, and dairy products tainted with odors and flavors imparted by weeds grazed by cattle.

Nonagricultural losses, though not so apparent, are nevertheless substantial. Railways, highways, and utility lines are kept clean from weeds and free from fire hazard at great expense. Weeds in cities, either in waste places or

1

vacant lots, or on lawns, are unsightly; many cause hay fever; some produce dermatitis in susceptible people, and others create fire hazards as the plants mature. Of the plants that cause hay fever the ragweeds alone affect the health of about 100 000 Canadians, entail medical expenses, and cause loss of time.

Under certain circumstances, some benefits may be derived from weeds, but these benefits should not be exaggerated. In drought areas weeds are often the only cover preventing soil from being blown; in moist areas they may prevent erosion on sloping banks. When plowed under, weeds add humus to soils. Weeds are useful as honey plants. Wildlife finds an important source of nutrition in weed seeds. Weeds are indifferent sources of forage, but during droughts they may be of great importance. Although many weeds are unsightly, some such as ox-eye daisy, orange hawkweed, blueweed, and goldenrods add considerable color to the landscape. The man who has to contend with these pests does not appreciate their esthetic value and he would be surprised to learn that some visitors from other countries find our common roadside weeds a striking part of the Canadian scene.

CHARACTERISTICS OF WEEDS

A weed is a plant that grows where man does not want it to grow, in grainfield, row crops, pastures, hayfields, lawns, and other disturbed habitats. Many plants designated as weeds could not survive or occur in their present abundance if these artificial habitats did not exist. In fact we are largely responsible for creating a suitable environment for the growth of the plants that we are most anxious to eliminate.

Weeds are often particularly suited to a certain type of habitat because of their life duration, habit of growth, or other characteristic. Weeds of row crops are usually annuals, whereas those of hayfields are mostly perennials. The leaves of many lawn weeds grow close to the ground. Weeds of pastures are sometimes not grazed by cattle because of spines or dense hairs. An unpleasant taste or odor serves a similar function.

Weeds cannot be separated from other plants by any particular characteristics. Weediness in more a matter of the extreme development of some habit of growth or seeding that is shared by many plants other than weeds. Weeds may be troublesome for one or several of the following reasons:

Many weeds are capable of growing under a wide range of climatic and soil conditions. There are exceptions: blueweed and hound's-tongue are confined to shallow limestone soil; wild carrot is a plant of Eastern Canada and British Columbia and it is absent from the Prairie Provinces.

A weed species often consists of a variety of forms with different physiological requirements, and, therefore, it can grow under the varied conditions presented by different habitats.

Many weeds produce abundant seed, far more than is required for the survival of the species, and these seeds are often easily disseminated either because of similar size of the seeds of the crop in which they grow or because they are equipped with barbs, as in blue bur, or tufts of hairs, as in dandelion, sow thistles, or milkweeds.

Seeds of many species such as the mustards are long-lived and they may present a problem for years, even though further seeding is prevented. As some seeds, such as wild oats, remain dormant throughout the winter, they defy early attempts at control.

A number of weeds, particularly the perennials, continue to grow after removal of the aboveground parts. The persistent perennials, toadflax for example, develop shoot buds on their roots and produce new shoots in spite of intensive cultivation to plow depth. Parts of the roots of some weed species are often spread during cultivation and they may establish the plant over a wider range than it formerly occupied.

Many weeds may grow and produce seeds under conditions adverse to crop plants. The vigorous growth of Russian thistle during drought years is an example.

Cutting the tops of weeds does not necessarily prevent seed reproduction, as some weeds can still mature seed if they are in flower at the time of cutting.

CONTROL

Control methods are under such rapid development that recommendations are soon out of date. Also, recommendations for weed control vary from one area to another, even within a province. For these reasons, the following discussion will be largely confined to the general control measures related to the different life durations of weeds.

Weeds are not uniform in life duration. Many are annual or winter annual, some are biennial, a large number are perennial. Any particular species may have annual and winter annual, or annual and biennial, or biennial and perennial forms. There is still much to be learned about the life duration of many weeds. Investigations of this kind are complicated by the influence of climatic factors on life duration.

Annuals complete their life cycles within 1 year and, although they may spread by the rooting of prostrate stems, as purslane, they have no means of survival from one year to the next except by seeds. Winter annuals germinate in the fall and pass through one winter as seedlings or rosettes; biennials germinate in the spring or early summer and survive the winter as rosettes. Both winter annuals and biennials complete their life cycles after producing seed in the second year. Therefore, control measures for these groups of weeds depend on the prevention of seed production by destroying the plants well before flowering and on the encouragement of germination by cultural practices. Rotation of crops and summerfallowing also help to control weeds. However, winter annuals and biennials may be attacked after harvest.

Simple perennials persist for more than 2 years. As these plants have no means of spreading other than by seeds, they may be treated in much the same way as annuals or biennials but deep root cutting may be necessary to completely eliminate them.

The weeds most difficult to control are the creeping or persistent perennials, which spread by means of seeds and by deep-rooted underground parts. It

3

is apparent that in spite of the application of severe control measures many of these plants are able not only to survive but to increase in lateral spread. Control depends on prevention of seed production, destruction of shoots and roots to plow depth, and also removal of top growth by mowing or by spraying with chemicals. It is rarely possible to completely eradicate these plants before the infested fields have to be used again for crops. The new chemicals help to prevent seed production while the land is in crop.

In recent years, the use of chemicals in agriculture has expanded remarkably and has revolutionized weed control practices. In fact, no other innovation in the history of agriculture has been accepted so rapidly and widely as the use of chemical weed killers. Such weed killers, known as herbicides, have now been devised for many control purposes, and probably even more effective and more specific herbicides will be discovered in the future.

Some herbicides, for example 2,4-D and MCPA, are so selective in action that they remove most broad-leaved weeds from grain crops, meadows, or lawns. At recommended rates, as little as 200-300 g of active chemical per hectare, weeds are killed or, in the case of perennials, prevented from flowering, without damage or with negligible damage to the crop. Herbicides of this type are applied annually to over 10 million hectares of grain in Western Canada. Herbicides that will prevent all plant growth for a year or longer are also available for use around industrial sites and similar areas.

In spite of the successful application of herbicides for weed control, other methods of control should not be considered as of secondary importance. Very often, a combination of tillage and spraying with herbicides will be more effective than either method alone. Some weeds respond very slightly to herbicides, and some crops are more sensitive to herbicides than the weeds are. In these instances, mechanical methods of control must be followed. Herbicidal sprays are not permanently effective in reclaiming grassland or in removing weeds from lawns if soil fertility is too low or drainage too poor for vigorous growth of grasses and clover. Under such conditions fertilization or drainage is needed to enable the grasses and clovers to compete successfully with weed and shrub seedlings. Grazing might also have to be adjusted to prevent deterioration of the grassland.

If you are in doubt about where to ask for advice and literature on weed control, the following lists of addresses, largely supplied by provincial weed specialists, will be helpful. Weed inspectors and agricultural representatives are not listed, but their wide knowledge of local conditions also provides an invaluable source of advice.

British Columbia

Agassiz. Agriculture Canada, Research Station.

Creston. Agriculture Canada, District Experiment Substation.

Kamloops. Agriculture Canada, Research Station.

New Westminster. British Columbia Department of Agriculture, Field Crops Branch.

Prince George. Agriculture Canada, Experimental Farm.

Saanichton. Agriculture Canada, Research Station.
Summerland. Agriculture Canada, Research Station.
Vancouver. Agronomy Department, Faculty of Agriculture, University of British Columbia.
Victoria. British Columbia Department of Agriculture, Field Crops Branch.

Alberta
Beaverlodge. Agriculture Canada, Research Station.
Edmonton. Alberta Department of Agriculture, Crop Clinic, Plant Industry Division.
Edmonton. Department of Plant Science, University of Alberta.
Lacombe. Agriculture Canada, Research Station.
Lethbridge. Agriculture Canada, Research Station.

Saskatchewan
Indian Head. Agriculture Canada, Experimental Farm.
Melfort. Agriculture Canada, Research Station.
Regina. Agriculture Canada, Research Station.
Regina. Saskatchewan Department of Agriculture, Plant Industry Branch.
Saskatoon. University of Saskatchewan.
Scott. Agriculture Canada, Experimental Farm.
Swift Current. Agriculture Canada, Research Station.

Manitoba
Brandon. Agriculture Canada, Research Station.
Morden. Agriculture Canada, Research Station.
Winnipeg. Manitoba Department of Agriculture, Legislative Buildings.
Winnipeg. Plant Science Department, University of Manitoba.

Ontario
Guelph. Department of Botany, Ontario Agricultural College, University of Guelph.
Harrow. Agriculture Canada, Research Station.
Kapuskasing. Agriculture Canada, Experimental Farm.
Kemptville. Kemptville Agricultural School.
Ottawa. Agriculture Canada, Scientific Information Section, Central Experimental Farm.
Toronto. Ontario Department of Agriculture and Food, Soils and Crops Branch, Parliament Buildings.
Vineland. Agriculture Canada, Research Station.

Quebec
L'Assomption. Agriculture Canada, Experimental Farm.
Lennoxville. Agriculture Canada, Research Station.
Montreal. Quebec Department of Agriculture and Colonization, Division of Crop Protection, Research Service, 201 Cremazie Boulevard East.
Montreal. Montreal Botanical Garden, 4101 East Sherbrooke Street.

Normandin. Agriculture Canada, Experimental Farm.
Quebec. Quebec Department of Agriculture and Colonization, Division of
    Crop Protection, Research Service.
La Pocatière. Agriculture Canada, Research Station.
St. Jean. Agriculture Canada, Research Station.

New Brunswick
Fredericton. Agriculture Canada, Research Station.
Fredericton. New Brunswick Department of Agriculture, Field Husbandry
    Branch.

Nova Scotia
Kentville. Agriculture Canada, Research Station.
Nappan. Agriculture Canada, Experimental Farm.
Truro. Botanist, Agricultural College.

Prince Edward Island
Charlottetown. Agriculture Canada, Research Station.
Charlottetown. Prince Edward Island Department of Agriculture.

Newfoundland
St. John's. Department of Mines and Resources, Agricultural Division.
St. John's West. Agriculture Canada, Research Station.

IDENTIFICATION OF WEEDS

Weeds are not easy to identify because of the large number of species, the variability within a single species, and the difficulties in recognizing weeds at different stages of growth. Manuals such as this can only assist in identification; much effort has to be made by the individual learning to identify weeds. Few people will be able to identify all the weeds growing on their farms and their only recourse is to send specimens of the unknown weeds to someone who has devoted more time to weed study than is possible on an average farm.

Too often, specimens are sent for identification without proper care being taken in collecting or packaging, and the material arrives at its destination in a condition that does not permit identification. The sender should collect the best possible material, preferably with flowers and seed-bearing parts and part of the underground system. Lack of a complete plant, however, should not deter anyone from sending a specimen. Unless the material is mailed promptly, it should be dried between papers and cardboards and placed under some weighty object. The packaging should be adequate to ensure safe arrival. When several plants are submitted, each specimen should bear a numbered tag corresponding to duplicates kept by the sender. Notes on habit, history of the infestation, abundance of the weed, and the place and habitat of the collection are often valuable for identification purposes and sometimes provide information that will later prove useful to the weed botanist in preparing publications or in giving advice.

Plants for identification may be submitted to provincial departments of agriculture, experimental farms and research stations of Agriculture Canada, the Line Elevators Farm Service, universities and agricultural colleges, agricultural representatives, weed inspectors, and the Biosystematics Research Centre, Agriculture Canada, Ottawa, Ontario. The list of addresses included in the previous section should also be consulted.

BOTANICAL TERMS

In this publication botanical terms have been kept to a minimum. Most of these terms are clarified by the drawings, or are defined in the text. The special terms used for the grasses are defined and illustrated on pages 12 and 13, and for the composites on pages 154 and 155; other terms that may not be familiar are italicized in the following remarks.

The illustration of a typical mustard flower, page 75, will remind readers of terms such as sepal, stamen, and pistil. The sepals are the usually green or leaflike whorl outside the petals and they are referred to collectively as the *calyx*. The petals as a group are referred to as the *corolla*. Within the petals are the stamens, the male or pollen-bearing organs. At the center of the flower is the *pistil* or female part of the flower consisting of one or more carpels. Each carpel is made up of a seed-containing lower part or *ovary*, which is surmounted by a style and stigma. The function of the stigma is to receive the pollen. *Inflorescence*, used occasionally in the text, means the flowering part of a plant.

Leaves without a stalk of any kind are referred to as *sessile*. If the leaf is completely divided into a number of divisions, as in poison-ivy, the individual divisions are *leaflets*. An *entire* leaf is one with an even margin that is not notched, toothed, or divided. *Glabrous*, as used in this manual, means without hairs.

SCIENTIFIC AND COMMON NAMES

Scientific names have been developed in accordance with systematic rules, and in general they make it possible to indicate precisely any plant or animal. Common names are based on general usage and often are misleading. Most plants have several common names, as will be seen in referring to the other names used for quack grass. Also, the same common name is sometimes applied to two or more species of plants; for example, creeping Charlie is used for two plants that are not even in the same family. Common names vary from language to language, and exchange of information on plants with other countries would be difficult if scientific names were not used.

The scientific names used in this book apply to families, genera, and species. A family ordinarily contains more than one genus. For instance, the genera (plural of genus) *Bromus, Agropyron*, and several others, are indicated in the following pages as belonging to the Gramineae family. A genus is a group of plants of higher rank than a species and usually includes several or many species that resemble each other in important structural characters. The scientific name of a plant consists of two parts: a genus name and a species name. The name *Bromus tectorum* L. refers to a species in the genus *Bromus*; the "L." following the scientific name in this example indicates that Linnaeus, the famous Swedish botanist, was responsible for naming this grass. In some cases the name first given to the plant has been changed and this fact is indicated by parentheses around the name or abbreviated name of the botanist who made the original description, for example, *Agropyron repens* (L.) Beauv. Linnaeus described this grass as *Triticum repens* and De Beauvois later transferred it to the genus *Agropyron*.

LITERATURE

The following brief selection will be of value to some readers:

Clark, G. H., and J. Fletcher. 1923. Farm weeds of Canada. Agriculture Canada. 192 p. Out of print, but copies are available in most libraries.

Kingsbury, John M. 1964. Poisonous plants of the United States and Canada. Prentice-Hall Inc., Englewood Cliffs, New Jersey. 626 p.

Kummer, Anna P. 1951. Weed seedlings. The University of Chicago Press, Chicago 37. 435 p. Illustrations and descriptions of the seedlings of 300 species, including all the common and noxious broad-leaved weeds found in Canada.

Muenscher, W. C. 1955. Weeds. The Macmillan Company, New York. 560 p. Descriptions, illustrations, and control measures for most of the weeds found in Canada.

Mulligan, G. A. 1987. Common weeds of Canada. A color illustrated guide to all the common weeds of Canada, available in most bookstores and libraries.

Wright, W. H. 1952-1953. Weed seeds. Agriculture Canada. Obtainable from the Queen's Printer, Ottawa. In 8 sets, each with 36 colored illustrations of seeds.

HORSETAIL FAMILY — EQUISETACEAE

FIELD HORSETAIL *Equisetum arvense* L.

Other names   Common horsetail, horse pipes, mare's tail, snake grass.

Description   Perennial, spreading by spores and by creeping rootstocks, which send out new shoots each year; stems annual, hollow, jointed, with sheaths at the joints; stems of two types: unbranched, 10-25 cm high, light brown, terminated by a cone, formed of shield-shaped stalked scales from which spores are produced; and branched, 20-25 cm high with whorls of 4-angled green branches arising at the joints, branches 10-15 cm long. Not a flowering plant; reproducing by means of spores that develop into a rarely seen minute sexual structure (gametophyte), from which the spore-producing plant develops. The unbranched stems ending in a cone appear by the middle of April and soon wither; the green branched stems, which last until frost, appear early in May.

Origin   Native to Canada.

Distribution in Canada   A common weed in all provinces, often a major weed problem, as in south central Manitoba and around Lesser Slave Lake in Alberta.

Habitats   Usually occurs on poorly drained soils; also in quite dry places, such as along railways. Hayfields, roadsides, and open woods.

Notes   Field horsetail is well known to be a poisonous plant. Horses, particularly young horses, are most seriously affected. Hay containing this weed is apparently more dangerous than pastures with horsetail. The deaths of many horses have been reported to veterinarians, but there is reason to believe that many more losses are unreported.

Similar plants   Field horsetail is one of a group of species with jointed and cone-bearing stems, characters that readily serve to distinguish horsetails from all other plants.

10

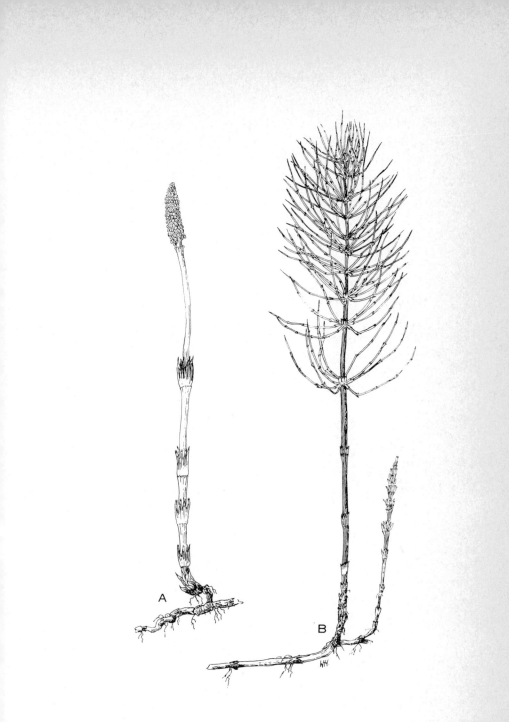

Field horsetail, **Equisetum arvense: A,** stem with spore-bearing cone; **B,** stem with branches.

GRASS FAMILY *Gramineae*

The grass family, because of its wide distribution, many species, and characters of growth, has greater economic importance than any other family of plants. Grasses, either as forage for grazing animals or as cereals, supply indirectly or directly the larger part of man's nutritional requirements. A number of grasses are serious weeds.

Grasses possess a unique structure that has acquired special descriptive terms, some of which cannot be avoided in describing a species. In the illustrations and the following descriptions the special terms used in the text will be clarified.

Grasses are annual or perennial. Many of the perennials possess rootstocks, which are creeping underground stems. Stems of grasses are usually hollow, except at the conspicuous nodes or joints. The leaves are solitary at the nodes and are arranged in two ranks on the stem. Each leaf has two distinct parts: the sheath, which forms a tube around the stem; and the blade, the long, narrow, nonclasping and usually flat part of the leaf. Except in barnyard grass, there is a tonguelike membranous or hairy outgrowth, the ligule, at the junction of sheath and blade. Auricles, clawlike projections at the base of the blade, are present in some grasses such as quack.

The grass flower lacks petals and sepals and consists of three stamens and a single ovary topped by two feathery stigmas. The flowers from species to species are so similar that identification must depend on the manner in which the flowers are grouped and on the structure, texture, and presence or absence of parts associated with the flowers.

Typically, each individual grass flower is enclosed between two scales or bracts, the flowering glumes. One of these, the lemma, is usually larger than the other, the palea, which it often encloses. The term floret is given to these units of lemma, palea, and the enclosed flower. Florets are arranged in spikelets. Typically, each spikelet consists of two empty glumes, within which are one or more florets. Spikelets are arranged in various ways. In some grasses, such as wheat or rye, the spikelets are sessile along the flowering stem and together they form a head. In still other grasses, as oats, the spikelets are arranged in a loose-branched inflorescence known as a panicle.

For practical purposes, the grass seed is here considered to be the kernel or caryopsis, together with its protective coverings, the lemma and palea. The kernel of most of the weed grasses remains within the lemma and palea even after threshing.

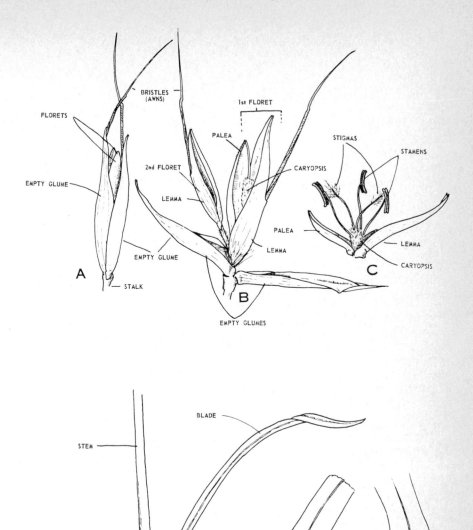

A, Spikelet of wild oats; B, spikelet with empty glumes bent back to show the florets (diagrammatic); C, a single floret (diagrammatic); D, grass leaf and part of stem; E, junction of leaf and sheath in a grass with a membranous ligule and with auricles; F, hairy ligule.

DOWNY BROME *Bromus tectorum* L.

Other names   Downy chess, early chess.

Description   Annual or winter annual; stems usually less than 6 dm high, often in large tufts; sheaths and leaf blades very hairy; spikelets numerous, hairy, narrow, on slender curved threadlike branches, each spikelet with 3-7 florets; lemmas with bristles (awns) emerging from below the tip; bristles rough, straight, 9-18 mm long; seed consisting of the caryopsis enclosed in the flowering scales (lemma and palea). Flowering early, seeds ripening from June to August.

Origin   Europe.

Distribution in Canada   Occurs in all provinces from New Brunswick to the Pacific Coast. Abundant only in southwestern Alberta and interior British Columbia.

Habitats   Overgrazed range, abandoned farmlands, around farm and ranch buildings, railroads, and roadsides. Thrives under dry conditions.

Similar plants   Downy brome is readily distinguished from other bromes found in Canada by its slender stems, hairy leaves, and the bristly spikelets on twisted branches.

Chess or cheat, *Bromus secalinus* L. (*C* in illustration), is a winter annual introduced from Eurasia. At the beginning of the century chess was considered to be a major noxious weed, but it is not so troublesome now. Chess is found more often in Ontario than in the other provinces, and is rarely found in the Prairie Provinces. The spikelets either lack bristles or have very short bristles and they contain more florets than the spikelets of downy brome.

Smooth brome, *Bromus inermis* Leyss., a valuable pasture and hay grass, can scarcely be regarded as a weed, although it often occurs on roadsides particularly in the western provinces. Japanese brome, *Bromus japonicus* Thunb., a European annual, is occasionally found as a weed from Quebec to Saskatchewan and it is rather abundant in southwestern Alberta.

Although brome grasses as a group are rather easily recognized, identification of the individual species requires careful study.

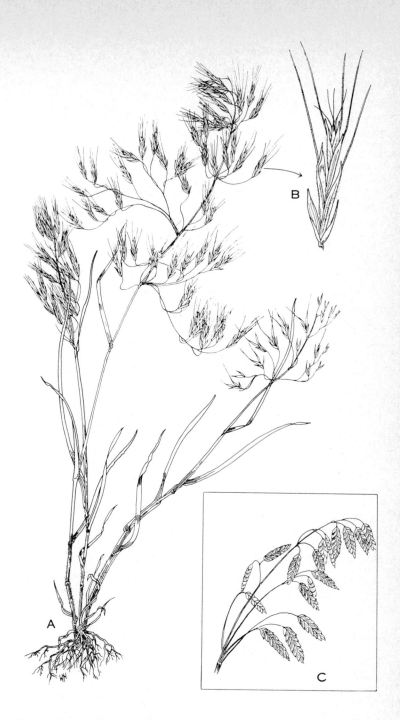

Downy brome, **Bromus tectorum: A,** plant; **B,** spikelet. Chess, **Bromus secalinus: C,** a group of spikelets.

GRASS FAMILY — GRAMINEAE

QUACK GRASS *Agropyron repens* (L.) Beauv.

Other names   Couch, quick, quitch, scutch, twitch grass.

Description   Perennial, spreading by seeds and rootstocks; rootstocks about 3 mm across, whitish or yellowish, cordlike, scaly, sharply pointed at the tip, abundant and widely creeping; flowering stems 3-12 dm high; leaves somewhat constricted below the tip, hairy on the inner surface, edges of leaf rough, sheaths of the lowest leaves usually hairy; auricles present; spikelets arranged in an unbranched, stiff, erect, and slender head, which is up to 15 cm long; spikelets usually just over 12 mm long, solitary and sessile in alternate notches of the flowering stem, the broader side against the flowering stem, each spikelet with 3-7 florets; empty glumes and lemmas often with bristles (awns); seed consisting of the caryopsis enclosed in the lemma and palea. Flowering toward the end of June.

Origin   Europe.

Distribution in Canada   Common in the agricultural areas of all provinces. Although abundant east of the Prairie Provinces at the beginning of the century, it was known only in a few localities in the west. Now common in the longer-settled regions of the west, although it may not yet have reached its maximum limits.

Habitats   Grasslands, cultivated fields, gardens, roadsides, and waste places.

Notes   The rootstocks of quack grass are very shallow where the infested land has been in sod for several years, but in cultivated soils they penetrate to a much greater depth and are much more vigorous.

Similar plants   The combination of matted, whitish rootstocks, auricles, hairy lower sheaths, and heads resembling a slender head of wheat are sufficient for separation of quack grass from most other grasses.

Quack grass resembles western wheat grass, *Agropyron smithii* Rydb., but the latter has bluish, rigid leaves that tend to roll in at the edges under dry conditions, while quack grass has lax, rarely bluish, leaves that always remain flat.

The rye grasses, *Lolium* spp., also possess slender spikes, but the spikelets are on edge to the flowering stem, whereas in quack grass they are arranged with the broader surface against the flowering stem.

Quack grass, **Agropyron repens: A,** plant; **B,** spikelet; **C,** seed from front and side; **D,** junction between leaf and sheath showing auricles.

PERSIAN DARNEL *Lolium persicum* Boiss. & Hoh.

Other names   Darnel. The scientific names *Lolium rigidum* Gaud. and *Lolium temulentum* L. have been incorrectly used for this species in North America. The first of these scientific names is properly applied to a forage species not grown in Canada and the second name refers to a plant introduced into Canada at different times but not known as a persistent weed.

Description   Bright-green annual; stems 15-50 cm high, branching from the base, all branches erect and close together; leaf blades rough on the inner surface and smooth on the outer; auricles usually present, spikelets edgewise to the stem, in two rows on opposite sides of the stem, sessile, each spikelet with 5-7 florets; empty glume as long as the spikelet or somewhat shorter, each spikelet with only 1 empty glume; caryopsis enclosed in the lemma and palea, the lemma bearing a bristle (awn) at the tip, bristle equal to the lemma in length.

Origin   Asia.

Distribution in Canada   Occurs in the western provinces and Ontario. Troublesome in the Peace River area, particularly in fields of commercial grasses grown for seed, and in recent years in south central Saskatchewan. Only known in Canada since 1923, it may become a more serious pest.

Similar plants   Persian darnel and other species of *Lolium* have spikelets placed on edge to the stem, whereas quack grass and its relatives have the broad side of the spikelet appressed to the stem.

Several European grasses related to Persian darnel are found in Canada. Italian rye grass, *Lolium multiflorum* Lam., is a serious weed only in the extreme southwest of British Columbia. It is a short-lived perennial with empty glumes shorter than the spikelet and with at least the upper lemmas with bristles.

Perennial rye grass, *Lolium perenne* L., is similar to Italian rye grass, but is less robust and its lemmas are usually without bristles. This short-lived perennial occasionally appears on lawns, sand ballast, and roadsides. It is used for grazing and forage on the Pacific Coast.

Darnel, *Lolium temulentum* L., differs from Persian darnel in that it is a taller plant, with plumper spikelets, and lemmas without bristles. (See note at the top of this page.)

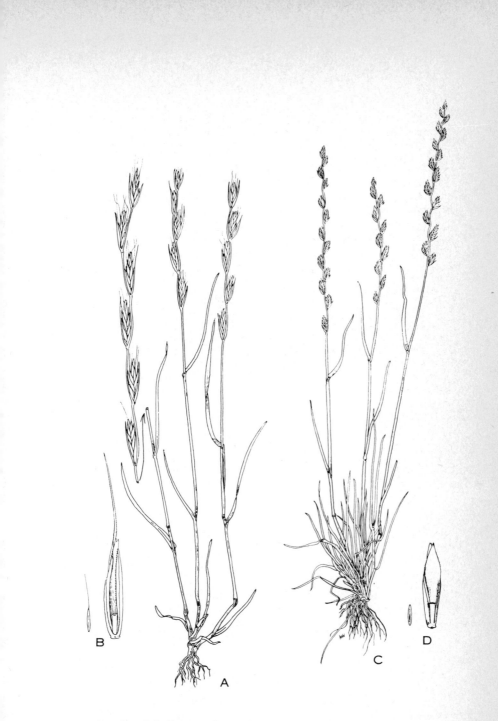

Persian darnel, **Lolium persicum: A,** plant; **B,** seed. Perennial rye grass,
Lolium perenne: C, plant; **D,** seed.

FOXTAIL BARLEY *Hordeum jubatum* L.

Other names   Skunk grass, squirrel-tail, wild barley.

Description   Perennial, spreading by seeds; stems 3-6 dm high, smooth, forming tufts; leaves grayish green; leaf blades rough, prominently ribbed, with long and short hairs on the inner surface, short hairs on the outer surface; sheaths ribbed but not rough, lower sheaths sometimes hairy; heads 5-12.5 cm long, at first tassel-like but at maturity nearly as broad as long; the head breaks up readily into 7-bristled clusters consisting of the 3 spikelets at each joint; the central spikelet of each cluster with a single floret and 3 bristles (awns), each outer spikelet represented by 2 bristles and lacking a floret; bristles green or reddish at early stages, becoming a shiny cream color at maturity, bristles to 75 mm long but may be much shorter. Flowering in July.

Origin   Western North America.

Distribution in Canada   Occurs in all provinces, but far less common east of the Great Lakes.

Habitats   Meadows, waste places, lawns, roadsides, ocean shores, borders of salt marshes, and dry saline depressions.

Notes   The slender flexible bristles easily become attached to animals that feed on hay contaminated with foxtail barley. The upward barbs of the bristles, then, are instrumental in forcing the sharp pointed base of the spikelets into the mouth, eyes, and skin, and may cause sores and sometimes blindness. The bristles rapidly work the spikelets to the back of the mouth and their barbs cause considerable irritation.

Similar plants   Foxtail barley is easily recognized by its nodding bushy head and it is not likely to be mistaken for other grasses.

Foxtail barley, **Hordeum jubatum.**

GRASS FAMILY — GRAMINEAE

WILD OATS *Avena fatua* L.

Description   Annual; flowering stems 6-12 dm high, erect; leaf blades long, flat, and broad; leaf sheaths and bases of leaf blades usually slightly hairy along margins especially in young plants; spikelets in a loose open panicle, drooping; each spikelet with 2 empty glumes within which are 2 or 3 florets, empty glumes longer than the florets; lemma hairy or glabrous, black, brown, gray, yellow, or white, with a bent twisted bristle (awn) about 3 cm long; seeds consisting of 2 flowering scales (lemma and palea) enclosing the caryopsis, all seeds with a slanting, circular, depressed scar (also called a sucker mouth) at the base, scar always with a circle of hairs. Flowering from the beginning of July.

Origin   Eurasia.

Distribution in Canada   Found in all provinces, but most troublesome in the grainfields of Western Canada.

Habitats   Grainfields, roadsides, and waste places.

Notes   Wild oats are well adapted as weeds of grainfields. The early ripening seeds remain as impurities in the crop seed. Delayed germination permits the plant to survive the winter as seed. Therefore, control measures late in the year are of little value.

Similar plants   The main difference between wild oats and cultivated oats is that the former sheds seeds readily but the cultivated oat varieties do not shed seeds. The seed in wild oats is attached to its minute stem by a joint and it easily falls off, retaining a circular scar at the base, whereas seeds of cultivated oats, dependent on fracture for separation from their stems, lack this circular scar. In addition each floret of the wild oats spikelet has a twisted bristle, whereas, in cultivated oats, bristles are either absent or confined to the lowest floret and then are usually straight; the panicle of wild oats is usually looser and spreads more widely than that of cultivated oats; germination of the seeds of wild oats is delayed, but the seeds of cultivated oats can germinate soon after maturity.

Oats that cannot be classed as wild or cultivated have been found at several localities in Western Canada. These off-types, including the so-called false wild oats, are very variable and some, at least, may have arisen from crosses between wild and cultivated oats.

Wild oats, **Avena fatua: A,** plant; **B,** spikelet; **C,** seed; **D,** seed after threshing.

SMOOTH CRAB GRASS *Digitaria ischaemum* (Schreb.) Muhl.

Other names   Finger grass, small crab grass.

Description   Annual; stems usually several, 7.5-45 cm high, usually less than 30 cm high, erect, becoming semiprostrate; leaf blades glabrous, flat, rarely over 4 mm wide and usually much narrower; sheaths smooth, upper sheaths always glabrous, those close to the ground often pubescent; ligule membranous, flower spikes 2-6, usually 4, appearing to arise from nearly the same point; spikelets about 2 mm long, each containing a single seed that is about as long as the spikelet; seed consisting of 2 shiny brown scales (lemma and palea) containing a yellow caryopsis. Flowering from mid-July to late September.

Origin   Eurasia.

Distribution in Canada   Found in all provinces except Newfoundland and Saskatchewan. An aggressive weed only in southern Ontario and Quebec.

Habitats   Lawns, gardens, roadsides, pastures, and waste places.

Notes   Crab grass is particularly troublesome in lawns, where its flowering stems, sprawling habit, purple color, and rapid growth in late summer are obnoxious. As an annual, it survives the winter as seed. Germination occurs only when the soil warms sufficiently, probably beginning about late May in Ontario and Quebec. Large numbers of seeds are produced before the frost kills the plants. Any attempts at eradication should include good lawn care and prevention of crab grass from maturing seed.

Similar plants   The only grass likely to be confused with small crab grass is large crab grass, *Digitaria sanguinalis* (L.) Scop., also of European origin. Large crab grass has about the same distribution as small crab grass, but it is found more often in agricultural crops and in gardens than in lawns.

Large crab grass has wider leaves, more numerous spikes, slightly larger spikelets, and it is generally taller than small crab grass. The hairy leaves of large crab grass permit easy distinction from small crab grass, which has glabrous leaves. The spikelet characters in the illustration show precise differences that can be seen under magnification: in tall crab grass the second glume is about half as long as the spikelet, and in small crab grass the second glume is equal in length to the spikelet.

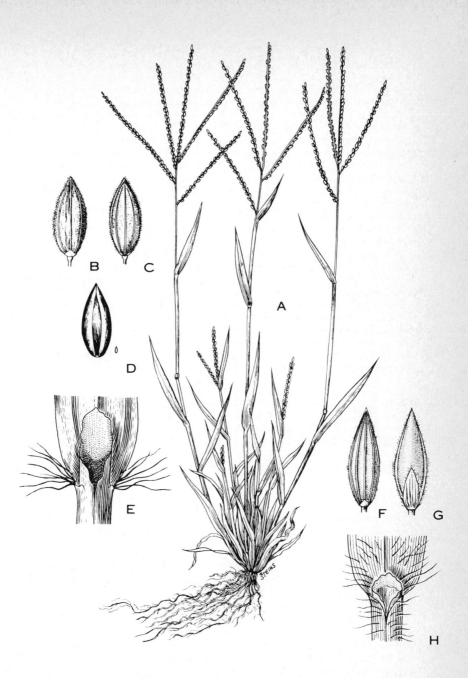

Smooth crab grass, **Digitaria ischaemum: A,** plant; **B,** front of spikelet with minute first glume at base; **C,** back of spikelet entirely covered by second glume; **D,** seed; **E,** junction of leaf blade and sheath. Large crab grass, **Digitaria sanguinalis: F,** front of spikelet with first glume at base; **G,** back of spikelet about half covered by second glume; **H,** junction of leaf blade and sheath.

GRASS FAMILY — GRAMINEAE

WITCH GRASS *Panicum capillare* L.

Other names   Capillary panic grass, common witch grass, old witch grass.

Description   Annual; stems usually several centimetres to 9 dm high, erect or spreading; leaf blades hairy on both surfaces, flat, leaves becoming progressively larger toward the top of the plant, the uppermost to 22 cm long; sheaths conspicuously hairy, strongly ribbed, with long hairs in the channels between the ribs; ligule a fringe of hairs; spikelets borne singly on slender stalks in a large, loose, open panicle; spikelets numerous, about 1.5 mm long, each containing one seed; seeds slightly shorter than the spikelet, consisting of 2 shiny, hard scales (lemma and palea) tightly enclosing the caryopsis. Flowering from July to September.

Origin   Native to North America.

Distribution in Canada   In every province except Newfoundland, and particularly troublesome in southern Ontario and southern Quebec.

Habitats   Gardens, cultivated fields, roadsides, waste places, and also in more natural locations such as on river and lake shores.

Notes   Witch grass germinates late, but grows vigorously and is well-developed by July. When mature, the main flowering stalk breaks off, and fruits or seeds are quite free in the spikelet and they are easily released.

Similar plants   A number of other panic grasses are closely related to witch grass, but this species is the one most often found in gardens and cultivated areas.

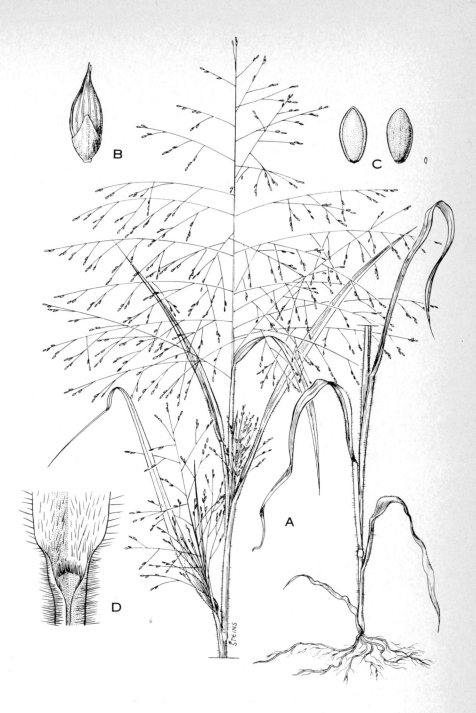

Witch grass, **Panicum capillare: A,** plant; **B,** spikelet; **C,** front and back view of seed; **D,** junction of leaf blade and sheath.

GRASS FAMILY — GRAMINEAE

BARNYARD GRASS *Echinochloa crusgalli* (L.) Beauv.

Other names   Cockspur grass, summer grass, water grass.

Description   A coarse glabrous annual; stems 3-12 dm high, usually several from the base, sometimes spreading over the ground; leaves long, broad, slightly roughened, spikelets numerous and crowded on branches of the flowering stem, upper branches often in a terminal clump; each spikelet with 1 floret and 2 empty glumes; empty glumes with scattered stiff hairs, one of the glumes often long-awned; threshing or rubbing removes the papery empty glumes and exposes the seed; seed consisting of a caryopsis tightly enclosed in the flowering scales (lemma and palea), seed about 3 mm long, white, yellowish, grayish, or brown, shining, rounded on one side and flattened on the other, the tip dark and wrinkled, and a ring of minute hairs on the rounded side just below the tip.

Origin   Introduced from Europe.

Habitats   Cultivated fields, gardens, barnyards, waste places, ditches, riverbanks, roadsides; usually on moist rich soil.

Distribution in Canada   Occurs in all eastern provinces, but rare from Manitoba west. The similar barnyard grass, *Echinochloa pungens,* abundant in Western Canada and less common in the east, may be absent in the provinces bordering on the Atlantic.

Similar plants   Most weed manuals include both barnyard grasses under the name *Echinochloa crusgalli.* These plants are dealt with separately here, as it is of some interest that *Echinochloa pungens* (Poir.) Rydb. has a very different distribution from the plant described and is native to North America. Originally this native species may have been confined to open ground such as eroding riverbanks, but with the increase in suitable habitats, it has now spread to roadsides, lawns, and cultivated fields. It is easily confused with *Echinochloa crusgalli,* and magnification is needed for recognition of the technical differences which are: seed lacking the ring of fine hairs below the tip, and the tip much firmer and of the same texture as the lustrous body of the lemma.

The barnyard grasses may be recognized readily because they are the only grasses in our area that are completely without a ligule.

Barnyard grass, **Echinochloa crusgalli: A,** plant; **B,** spikelet; **C,** spikelet with empty glumes removed exposing the seed.

GREEN FOXTAIL *Setaria viridis* (L.) Beauv.

Other names   Bottle grass, green bristle grass, wild millet.

Description   Annual; stems single or several from the base, 75 mm to 9 dm high, usually about 4 dm; leaves narrowing to a fine point at the tip, glabrous, rough, flat; sheaths smooth, margins short, hairy; ligule a fringe of short hairs; spikelets densely grouped in a green, soft, bristly, elongated head, which resembles a slender bottle brush, heads 2.5-10 cm long; the green and finally straw-colored spikelets are exceeded by the bristles, which arise below them and remain on the plant after the spikelet has shattered from its minute cuplike base; threshing or rubbing removes the papery, empty glumes from the spikelet and exposes the single seed; seed consisting of 2 hard, minutely warty, shiny scales (lemma and palea), which tightly enclose the caryopsis; seed about 1.5 mm long and whitish, pale yellow, or purplish, depending on stage of maturity. Flowering from July to September.

Origin   Europe.

Distribution in Canada   Found in all provinces and becoming a major pest in Western Canada.

Habitats   Grainfields, gardens, roadsides, and waste places.

Notes   Seeds of green foxtail germinate largely between May 15 and June 15, so that early spring and late summer cultivation have little effect on control.

Similar plants   Yellow foxtail, *Setaria glauca* (L.) Beauv., introduced from Europe, appears to be confined, as an abundant weed, to the coastal area of British Columbia and to Eastern Canada east of the Great Lakes. It differs from green foxtail in the following respects: long twisted hairs on the upper surface of the leaf blade near the base; leaf sheaths without hairy margins; heads with fewer spikelets; seeds conspicuously cross-wrinkled and much larger, nearly 3 mm long. The bristles of yellow foxtail appear faintly yellowish brown at maturity, whereas the bristles of green foxtail are straw-colored.

Bristly foxtail, *Setaria verticillata* (L.) Beauv., introduced from Europe, is almost exclusively a weed of gardens and townsites. It can be recognized by the downwardly directed barbs on its bristles. All other foxtails have barbs directed upward.

Giant foxtail, *Setaria faberii* Herrm., an Asiatic grass, now spreading rapidly in the north central and eastern United States, may eventually reach Canada. It is a coarse annual, 15-18 dm high, with an arching head up to 18 cm long, and with the leaves hairy over the upper surface.

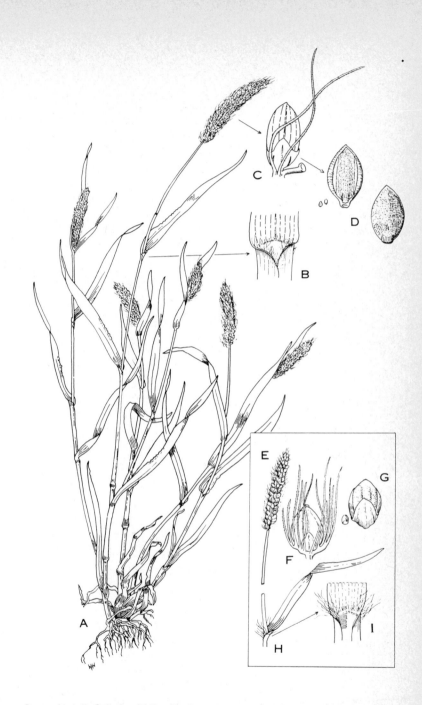

Green foxtail, **Setaria viridis: A,** plant; **B,** junction of leaf blade and sheath showing ligule; **C,** spikelet; **D,** seed. Yellow foxtail, **Setaria glauca: E,** head; **F,** spikelet; **G,** seed; **H,** leaf blade showing long hairs at base; **I,** junction of leaf blade and sheath.

CURLED DOCK *Rumex crispus* L.

Other names   Sour dock, yellow dock.

Description   Perennial, spreading by seeds; taproot stout, yellowish when cut across; stem usually at least 9 dm high, ridged; leaves dark green, 8-30 cm long, much waved and crisped along margins; above each leaf a delicate sheath surrounds the stem; flowers small and spread along the upper part of the stem and branches in clusters, at maturity forming a dense brown inflorescence, flower stalks with a swollen joint; the three inner sepals (valves) quite prominent, somewhat heart-shaped, each bearing a tubercle, of which one is larger than the other two; seeds reddish brown, about 2 mm long, shiny, 3-sided, and therefore triangular in cross section. Flowering from mid-June and producing seeds from July to September.

Origin   Eurasia.

Distribution in Canada   Occurs in all provinces, and most abundant in the eastern provinces.

Habitats   Meadows, pastures, roadsides, fencerows, and waste places.

Similar plants   Four species of dock often confused with curled dock are shown on page 35.

None of these four species has all the following characters that distinguish curled dock: leaves narrowed at the base; leaf margins crumpled and wavy; valves with entire margins; and valves with tubercles. Leaves of *Rumex longifolius,* long-leaved dock, are broader and not narrowed at the base and its valves lack tubercles. Leaves of field dock, *Rumex pseudonatronatus,* are not as wavy and the valves either lack tubercles or one may have a slight swelling. In serrate-valved dock, *Rumex stenophyllus,* and in broad-leaved dock, *Rumex obtusifolius,* the valve margins are not entire and in the latter species the leaves are heart-shaped at the base.

Broad-leaved dock often hybridizes with curled dock. The teeth on the valves of the hybrid are shorter than the teeth on the valves of broad-leaved dock, and therefore the hybrid has been mistaken in Europe and North America for serrate-valved dock, *Rumex stenophyllus.* The hybrid has short stubby hairs on the leaf petiole and on the veins of the lower leaf surface, and only one of its three valves has a well-developed tubercle. Serrate-valved dock has narrower leaves and no hairiness on leaves or petioles, and each of its three valves has a well-developed tubercle. The hybrid has been found in Newfoundland, Nova Scotia, Quebec, Ontario, and British Columbia, and serrate-valved dock is largely confined to the Prairie Provinces.

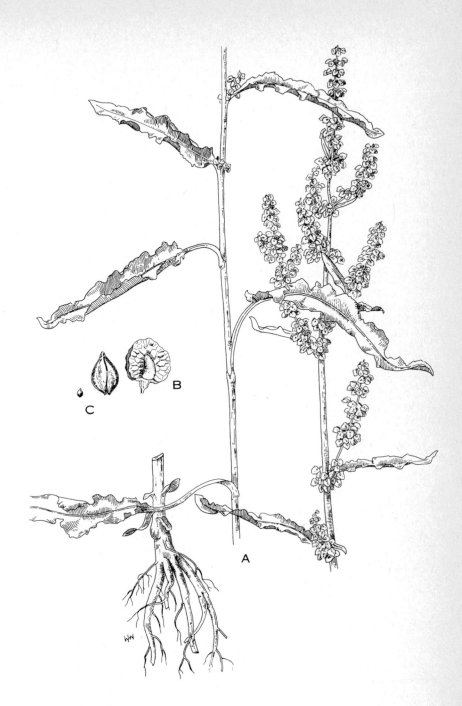

Curled dock, **Rumex crispus: A,** plant; **B,** sepal (valve) with tubercle; **C,** seed.

LONG-LEAVED DOCK *Rumex longifolius* DC.
Other names   Formerly known by the scientific name *Rumex domesticus* Hartm.
Description   Stout perennial; stems usually about 7 dm high; leaves slightly wavy on the margins, about three times as long as broad, oblong, usually broad at the base; valves roundish, about 6 mm long, without tubercles; seeds brown, 3 mm long, narrowed to the tip.
Notes   This Eurasiatic weed is known from all provinces except Alberta, but is most common in Quebec and Ontario. Long-leaved dock has been mistaken for patience dock, *Rumex patientia* L., a much taller plant, to 18 dm, with larger valves, of which one has a plump tubercle. Patience dock is rare in Canada and is only known from Ontario.

FIELD DOCK *Rumex pseudonatronatus* Borbas
Other names   Formerly known by the scientific name *Rumex fennicus* (Murb.) Murb.
Description   Slender perennial; stems to 15 dm high; leaves narrow, to 15 times as long as broad, leaf base narrowed to the petiole; valves about 1.5 mm long, without tubercles, but one valve possibly with a slight swelling; seeds brown, 2.5 mm long or less.
Notes   This Eurasiatic weed, first collected in Canada in 1938, has become much more abundant than curled dock in the Prairie Provinces. Field dock is common in Saskatchewan and Manitoba, but it is also known to occur in Ontario, Alberta, and British Columbia. There is an isolated record for the Yukon.

SERRATE-VALVED DOCK *Rumex stenophyllus* Ledeb.
Description   Perennial; stems usually 9-12 dm high; leaves only slightly wavy on the margins, narrowed toward both ends, without stubby hairs on midribs and petioles; valves about 4 mm long, margins with several, short, coarse teeth, each valve with a well-developed tubercle; seeds brown, 2.5 mm long.
Notes   This dock, native to Russia, has spread into Europe and more recently to North America. It was first collected in Canada in 1950, and is now known as a common weed in Saskatchewan. It is also found in Manitoba, in Alberta, and at Rimouski in Quebec. It is often confused with a hybrid between two other species (see page 32).

BROAD-LEAVED DOCK *Rumex obtusifolius* L.
Description   Perennial; stems to about 9 dm high; leaves broad and with heart-shaped bases, petioles and veins on leaf undersurfaces with short blunt whitish hairs; valves 5 mm long, triangular in outline, margin with several long teeth, usually only one valve with a prominent tubercle; seeds brown, 2 mm long.
Notes   Broad-leaved dock, introduced from Europe, occurs from Newfoundland to Ontario and in British Columbia. There is no evidence that it occurs in the Prairie Provinces.

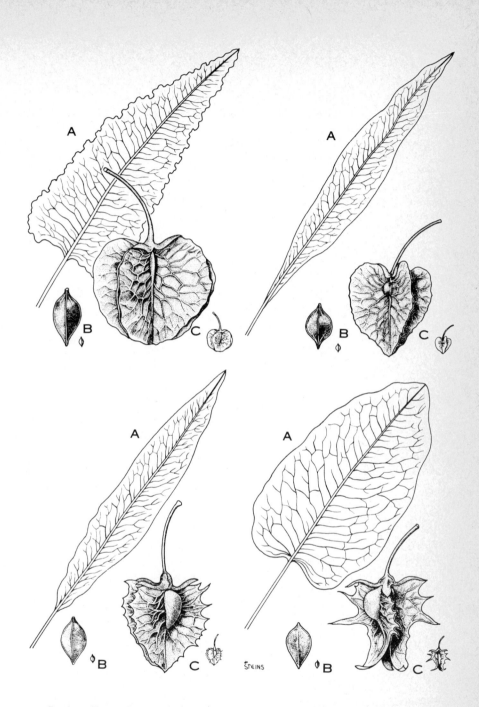

Docks. **Upper left:** Long-leaved dock, **Rumex longifolius. Upper right:** Field dock, **Rumex pseudonatronatus. Lower left:** Serrate-valved dock, **Rumex stenophyllus. Lower right:** Broad-leaved dock, **Rumex obtusifolius.** In all drawings: **A,** basal leaf; **B,** seed; **C,** valves, enlarged and natural size.

SHEEP SORREL *Rumex acetosella* L.

Other names   Field sorrel, horse sorrel, red sorrel, sourgrass.

Description   Perennial, spreading by seeds and much-branched slender root-stocks; stems 15-50 cm high, numerous and wiry; leaves green and sour in taste, alternate, the lower leaves borne on long stalks and usually with a pair of spreading lobes at the base, the uppermost leaves without stalks or lobes; inflorescences terminating the stems; flowers small, either seed-producing (female) or pollen-producing (male), each sex on separate plants; flowers without petals but with 6 sepals in 2 circles of 3, the inner sepals much more prominent than the outer; in the female, flowers tightly enclosing the seed, flower stalks without joints; seeds triangular, 3-sided in cross section, reddish brown, surface smooth and shiny, about 1.3 mm long and nearly as broad. Flowering throughout the summer.

Origin   Eurasia.

Distribution in Canada   Rare in the Prairie Provinces, common in southern British Columbia and Eastern Canada. Occurs in all provinces.

Habitats   Meadows, pastures, roadsides, often on sandy soils. Sheep sorrel is often found on acid soils, but, since it also occurs on alkaline soils, it is not a good indicator of acid conditions. Its presence is far more likely to be an indication of soil impoverishment. Some soils contain a large number of seeds of sheep sorrel, and, if the clover seeding fails, the resulting stand of sorrel is often mistakenly thought to be due to contamination of the agricultural seed.

Similar plants   One or other of three species of garden sorrel, *Rumex acetosa* L., *Rumex rugosus* Campd., and *Rumex thyrsiflorus* Fingerh., introduced from Europe, is found in all provinces. *R. thyrsiflorus* is particularly abundant in Quebec along the St. Lawrence River. These species are taller plants, about 9 dm high, and have larger leaves than sheep sorrel. A less evident difference is found in the stalks of the flowers, which have a distinct joint that is lacking in sheep sorrel.

Sheep sorrel, **Rumex acetosella: A,** plant; **B,** seed enclosed in sepals and seed with sepals removed.

PROSTRATE KNOTWEED *Polygonum aviculare* L.

Other names   Doorweed, knotgrass, matgrass.

Description   Annual, spreading by seed; stems freely branched from the base, prostrate to occasionally semierect; leaves up to 5 cm long but usually smaller, 3-5 times as long as broad, broadest near the middle and narrowed towards the tip and base; a tubular, papery sheath at the base of each leaf encircles the stem; flowers small, one to few in the leaf axils, without petals but with 5 small greenish to purplish sepals; sepals tightly enclosing a single seed; seed dull brown, 3-sided, minutely roughened, about 2 mm long. Flowering June to September.

Origin   Introduced from Eurasia.

Distribution in Canada   Found in all provinces.

Habitats   Most common on trampled land around habitations, but also roadsides, waste places, and occasionally in cultivated fields.

Similar plants   Several closely related plants in the genus *Polygonum* resemble prostrate knotweed. Most of these are native plants and of little importance as weeds. The differences between these plants and prostrate knotweed are very minute and not well understood, even by botanists.

One of the most distinctive and widespread native species is striate knotweed, *Polygonum achoreum* Blake. This knotweed, particularly common on roadsides, is easily distinguished from prostrate knotweed by its more erect habit, coarser stems, broader, more rounded leaves, hooded sepals, and smooth, olive-colored seeds. Striate knotweed is found in all provinces except Newfoundland and Prince Edward Island.

Although knotweeds and smartweeds are in the same genus, the flowers of knotweeds are in the axils of leaves, whereas in smartweeds the flowers are grouped at the ends of stems.

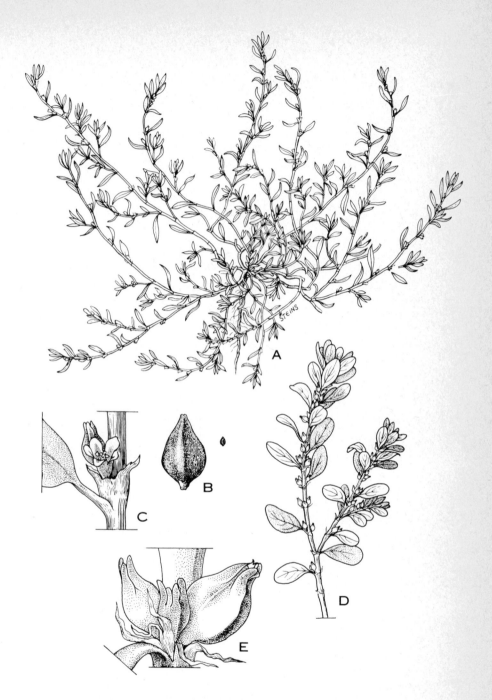

Prostrate knotweed, **Polygonum aviculare: A,** plant; **B,** seed; **C,** flowers
emerging from sheath at base of leaf. Striate knotweed, **Polygonum
achoreum: D,** plant; **E,** flowers emerging from sheath at base of leaf.

BUCKWHEAT FAMILY — POLYGONACEAE

GREEN SMARTWEED *Polygonum scabrum* Moench

Description   Annual; stems 3-9 dm high, usually branched, smooth except for gummy unstalked glands below the spikes; leaves narrowed to tip and base, often with a dark blotch at the center, the basal leaves usually with white cobwebby hairs on the lower surface, upper leaves often without hairs and with round yellowish-brown glandular dots on the lower surface; at the base of the leaf and encircling the stem is a tubular sheath with a fringe of minute bristles not evident without magnification (see illustration); flowers green, grouped in blunt, thick, erect spikes; petals lacking, sepals usually with raised veins, sepals not forming a beak over the tip of the seed; seeds about 2.5 mm broad, roundish, abruptly pointed at the tip, flattened, both surfaces somewhat hollowed, shining, black.

Origin   Europe.

Distribution in Canada   Found in every province but more abundant in the Maritime Provinces.

Habitats   Grainfields, cultivated fields, roadsides, and waste places.

Similar plants   Lady's-thumb, *Polygonum persicaria* L., introduced from Europe, occurs in all provinces. Green smartweed is much commoner than lady's-thumb in the Prairie Provinces and in eastern Quebec, and the provinces bordering on the Atlantic. Lady's-thumb differs from green smartweed in the following ways: spikes usually pinkish, upper margins of sheaths with long hairs that can be seen without magnification, leaves without evident glandular dots, seeds either three-sided or flattened but not hollowed on the surfaces.

The black blotch on the leaf of lady's-thumb cannot be relied on for identification of this species because the leaves of both pale and green smartweed often have this blotch.

Pale smartweed or pale persicaria, *Polygonum lapathifolium* L., is native to North America and Europe. It occurs in all provinces, and in some areas, particularly in the Prairie Provinces, it is the most abundant and persistent smartweed. Pale smartweed differs from the other species mentioned here by its elongate and nodding spikes, usually pinkish and rarely white.

Green smartweed, **Polygonum scabrum: A,** plant; **B,** leaf base and sheath; **C,** seed. Lady's-thumb, **Polygonum persicaria: D,** 3-sided seed; **E,** flattened seed.

BUCKWHEAT FAMILY — POLYGONACEAE

WILD BUCKWHEAT *Polygonum convolvulus* L.

Other names   Black bindweed, climbing bindweed, corn bindweed. In Europe placed in a separate genus as *Bilderdykia convolvulus* (L.) Dumort.

Description   Annual; stems slender, trailing on the ground or twining about other plants, freely branched from the base, usually greenish or olive green; leaves alternate, smooth, pointed at the tip, somewhat heart-shaped with widely separated basal lobes; flowers small, greenish, drooping, short-stemmed, in clusters in the leaf axils or in longer slender clusters at the tips of branches, without petals but with 5 sepals; sepals tightly enclosing the single seed, becoming brown when fully ripe; seeds dull black, 3-sided, minutely roughened, 3 mm long or slightly longer, the seed as found in crop seeds may be covered by the sepals. Flowering in late June and July.

Origin   Europe.

Distribution in Canada   Common in the agricultural areas of all provinces, but more abundant in the west than in the east.

Habitats   Grainfields, cultivated fields, gardens, along railroad tracks and roads, waste places, and occasionally in open woods or at wood margins.

Similar plants   Wild buckwheat and field bindweed (see page 136) have twining, greenish stems and leaves of similar shape and may be mistaken for each other before flowering occurs unless attention is paid to the root systems, annual in wild buckwheat and perennial in field bindweed. Wild buckwheat, in common with other species of the buckwheat family, has sheaths at the leaf bases encircling the stem. These sheaths are lacking in field bindweed. When in flower, these plants are easily distinguished, because wild buckwheat has small greenish flowers and field bindweed has large pink or whitish flowers.

Two native species of *Polygonum* resemble wild buckwheat, although they are not field weeds and are only likely to be found at the edges of woods, in clearings, and on rocky slopes. Fringed wild buckwheat, *Polygonum cilinode* Michx., differs from all other species of *Polygonum* in having bristles at the base of the sheath, and from wild buckwheat in its perennial habit, reddish stems, leaves with a narrower space between the basal lobes, and smooth and shiny seeds. Climbing false buckwheat, *Polygonum scandens* L., has a strongly winged calyx, long-stemmed flowers, and smooth, shiny seeds. Both plants occur in the eastern half of the continent.

Wild buckwheat, **Polygonum convolvulus: A,** plant; **B,** seedling; **C,** part of flowering stem at maturity; **D,** seed covered by sepals and also partially exposed.

BUCKWHEAT FAMILY — POLYGONACEAE

TARTARY BUCKWHEAT *Fagopyrum tataricum* (L.) Gaertn.

Description   Annual, spreading by seed; stems erect, usually green; leaves light green, mostly broader than long, triangular heart-shaped, with wide-spreading lobes at the base, lower leaves on long stalks, upper leaves short stalked; flowers greenish, small, clustered at the ends of stems and in leaf axils; seed strongly protruding from the sepals, rather sharply 3-sided and edged near the tip, in the lower part (toward the sepals) wrinkled and notched on the edges, brownish or dark gray, rough and dull, 4 mm long or somewhat longer. Flowering in early summer and maturing seed over a rather extended period.

Origin   Asia.

Distribution in Canada   The distribution of Tartary buckwheat is not well understood. It occurs in the eastern provinces; but it is a serious weed only in north central Alberta, where it is spreading rapidly, and in Manitoba, north of Riding Mountain National Park. Tartary buckwheat probably will spread more widely in Western Canada.

Tartary buckwheat has been grown extensively as a crop plant in Eastern Canada, particularly in New Brunswick, but is probably less often cultivated now.

Habitats   Grainfields, waste places, and roadsides.

Notes   Seeds of Tartary buckwheat possess a considerable dormancy. Storage of seeds under dry conditions for 2 months or more permits rapid germination. The seed is difficult to remove from wheat.

Similar plants   Buckwheat, *Fagopyrum sagittatum* Gilib., is similar to Tartary buckwheat in appearance. It differs from Tartary buckwheat in having somewhat reddish stems, larger white or reddish flowers, thicker clusters of flowers but, above all, in having smoother and more sharply triangular seeds. Tartary buckwheat is more resistant to drought and frost than buckwheat, and it matures seeds earlier.

Both these plants have seeds strongly protruding from the sepals and this character permits ready distinction from wild buckwheat. The buckwheats are erect plants, while wild buckwheat has a prostrate or twining habit. The smooth, black, small seeds of wild buckwheat are readily distinguished from the larger, grayish, roughened seeds of Tartary buckwheat.

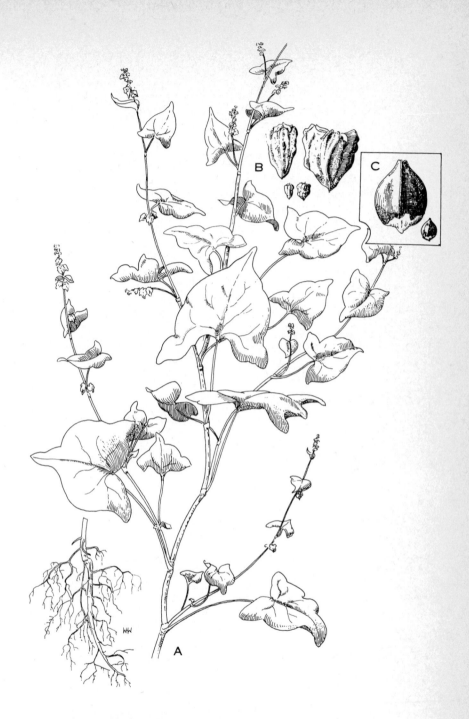

Tartary buckwheat, **Fagopyrum tataricum: A,** plant; **B,** seeds. Buckwheat,
Fagopyrum sagittatum: C, seed.

KOCHIA *Kochia scoparia* (L.) Schrader

Description   Annual, spreading by seed; stems erect, 3-18 dm high, much branched; leaves alternate, stalkless, long and narrow, tapering to a point at the tip, margins entire, leaves of main stem larger than those of the branches, becoming progressively smaller along the branches towards the tips, sometimes turning purplish in the autumn; flowers small, stalkless, inconspicuous, without petals, one to several flowers in the axils of leaves on the branches, each of the 5 sepals with a thickish wing, flowers sometimes surrounded by a cluster of long hairs, flowers containing a single seed that is covered by a thin papery envelope; seeds about 2 mm long, dull brown, nearly oval, flattened, and with a groove on both sides. Flowering from July to September.

Origin   Introduced from Eurasia.

Distribution in Canada   Kochia is known from Nova Scotia, Quebec, Ontario, Manitoba, Saskatchewan, Alberta, and British Columbia. Common only in the Prairie Provinces and the Okanagan Valley of British Columbia.

Habitats   Waste places and roadsides; becoming a problem in cultivated fields in the Prairie Provinces.

Notes   A form of kochia, burning bush or summer cypress, is grown in gardens for its symmetrical oval or columnar habit and brilliant purple-red color in autumn. This form occasionally escapes from cultivation, although it does not appear to persist. There is always the possibility that it may revert to the coarsely branched broader-leaved weedy plant. However, no definite information is available.

Similar plants   Five-hooked bassia, *Bassia hyssopifolia* (Pall.) Ktze., resembles kochia. The two plants can be readily separated by the presence of a long hooked prickle on each of the sepals of five-hooked bassia; the sepals of kochia lack these prickles.

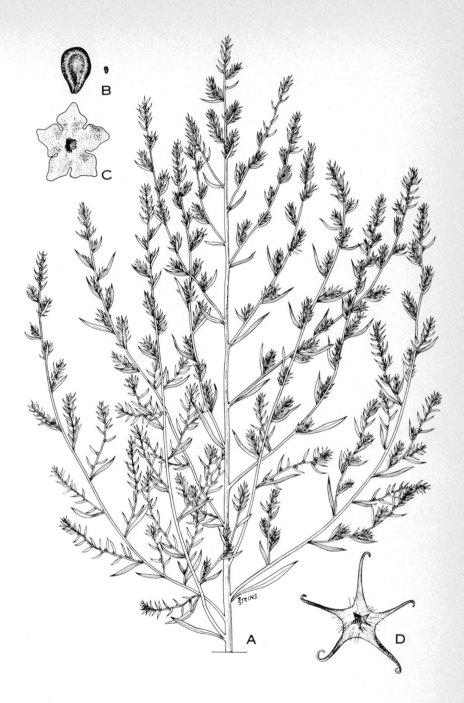

Kochia, **Kochia scoparia: A,** plant; **B,** seed; **C,** flower from above showing the five thick sepals. Five-hooked bassia, **Bassia hyssopifolia: D,** flower from above.

47

GOOSEFOOT FAMILY — CHENOPODIACEAE

LAMB'S-QUARTERS *Chenopodium album* L.

Other names   Fat-hen, pigweed, white goosefoot.

Description   Annual, spreading by seeds; stems 3-18 dm high, branched, ridged, green or sometimes purple-striped; leaves alternate, stalked, very variable in shape, somewhat triangular, margins coarsely toothed but sometimes nearly entire, lower surface grayish green and covered with mealy particles; flowers small, inconspicuous, greenish, densely crowded in the leaf axils and at the stem tips, without petals, at maturity each flower containing a single seed; seeds about 1.3 mm across, shiny black, flattened, nearly circular in outline, often covered by a thin, white, papery envelope (pericarp). Flowering from June to September.

Origin   Europe.

Distribution in Canada   Lamb's-quarters is one of the most abundant weeds in the agricultural areas of Canada.

Habitats   Cultivated land, grainfields, gardens, roadsides, and waste places.

Similar plants   There are several widespread species that are hard to separate from lamb's-quarters.

Maple-leaved goosefoot, *Chenopodium gigantospermum* Aellen, a North American native, occurs from New Brunswick to British Columbia as a weed of fields and waste places. It is similar to lamb's-quarters except for thin green leaves (see illustration) and larger seeds.

Oak-leaved goosefoot, *Chenopodium glaucum* L., as found in all provinces of Eastern Canada is largely of European origin, while in the western provinces it is native. This goosefoot is a shorter plant than lamb's-quarters, and has smaller leaves, which are white on the under surfaces.

Strawberry blite, *Chenopodium capitatum* (L.) Asch., is a native plant found from Quebec to British Columbia. It is the most distinctive of this group of plants as its clusters of flowers become fleshy and red when mature.

Net-seeded lamb's-quarters, *Chenopodium berlandieri* Moq. var. *zschackei* (Murr.) Murr., a native plant found from Ontario to British Columbia, is probably more abundant than lamb's-quarters in the Prairie Provinces. It can be distinguished from lamb's-quarters by its thicker leaves and striking honeycombed markings on its seeds and pericarps.

Late-flowering goosefoot, *Chenopodium strictum* Roth var. *glaucophyllum* (Aellen) Wahl, separable from other *Chenopodium* species by flowering in early September and by shallow teeth on lower leaves, is native to North America. This late-flowering species occurs from Quebec to British Columbia.

Lamb's-quarters and its relatives, *Chenopodium* spp., have alternate leaves and can be separated at any stage from the superficially similar species of the genus *Atriplex,* which have at least the first three or four pairs of leaves opposite.

48

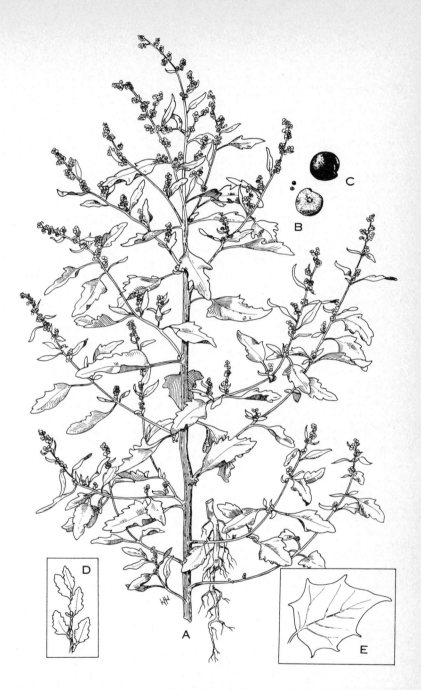

Lamb's-quarters, **Chenopodium album: A,** plant; **B,** seed covered by papery envelope; **C,** seed. Oak-leaved goosefoot, **Chenopodium glaucum: D,** part of shoot. Maple-leaved goosefoot, **Chenopodium gigantospermum: E,** single leaf.

GOOSEFOOT FAMILY — CHENOPODIACEAE

RUSSIAN PIGWEED *Axyris amaranthoides* L.

Description   Annual; stems erect, to 12 dm high, usually much branched, light green becoming white at maturity; leaves alternate, pale green, not toothed or only rarely with a few obscure teeth, the lower surface with numerous star-shaped hairs, the upper surface with fewer hairs or glabrous in older leaves, leaves of the main stem much larger than those of the flowering branches; flowers of two kinds on the same plant: male flowers in elongate slender clusters at the ends of branches, yellowish, disappearing after flowering; female flowers in the axils of bractlike leaves, without petals but with 3 or 4 papery sepals, sepals longer than the seeds; seeds of two types: with a 2-lobed wing at the tip, oval, 3 mm long, dark brown, under magnification appearing mottled from minute gray wrinkles scattered over the seed surface and arranged lengthwise; and wingless, roundish, 2 mm long, smooth-surfaced, evenly olive-gray. Flowering from June to August.

Origin   Asia.

Distribution in Canada   Occurs in all provinces except Newfoundland. One of the most abundant weeds in the Prairie Provinces, where it extends to the northern limits of agriculture; not yet common in the east. First noted in 1886 by the roadside at Headingly, Manitoba.

Habitats   Grainfields, roadsides, gardens, farmyards, manure piles, waste ground, and along railways.

Notes   Experiments at Swift Current have shown that the seeds without wings germinate much less readily than those with wings and apparently may persist for a long period as a source of contamination in soils.

Similar plants   There is little danger of confusing Russian pigweed with other plants. In young stages of growth the leaves of Russian pigweed are not unlike those of lamb's-quarters but they have star-shaped hairs and are not mealy on the lower surface.

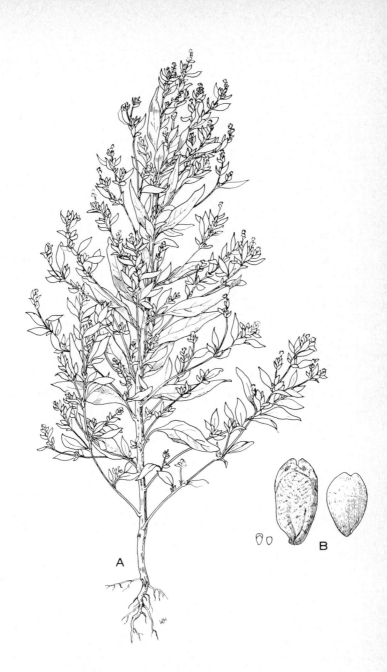

Russian pigweed, **Axyris amaranthoides: A,** plant; **B,** winged and wingless
seeds.

GOOSEFOOT FAMILY — CHENOPODIACEAE

RUSSIAN-THISTLE *Salsola pestifer* Nels.

Other names   Russian cactus, Russian tumbleweed. The scientific name *Salsola kali* L. var. *tenuifolia* Tausch is sometimes applied to this plant.

Description   Annual; stems from several centimetres to nearly 12 dm high, often reddish-striped; leaves alternate, 2-5 cm long, narrow, flattened on the upper side, tipped by a sharp point; flowers in the leaf axils, inconspicuous, with 5 papery sepals winged at maturity and often twisted at the tips to form a cone over the single seed; each flower in a cup-shaped depression formed by a leaf base and 2 spiny-tipped stiff bracts, bracts about 6 mm long; the seed falls from the plant with the winged sepals attached, seeds top-shaped, the broader end somewhat hollowed and with a central protuberance, nearly 2 mm across, dull gray, seed coat so transparent that the spreading coiled embryo can be seen. Flowering from the beginning of July; maturing seed by mid-August.

Origin   Eurasia.

Distribution in Canada   Occurs in all provinces except Newfoundland. Abundant in the drier parts of the Prairie Provinces. Far less common in Eastern Canada, where it is almost exclusively a railway and roadside weed.

Habitats   Cultivated fields, roadsides, waste places, fencerows, and along railroad tracks. Common on the drier soils.

Similar plants   Russian-thistle has distinctive characters at all stages of growth. The young plant is soft and tender and bears elongated leaves that fall as the seeds mature. At maturity the plants become hard and rigid, and the stiff, spiny bracts are the most prominent feature. Late in the year the nearly spherical bushy top breaks away at the ground line and is rolled readily by the wind, dropping seeds in its path. Russian-thistle should not be confused with another tumbleweed, *Amaranthus albus,* which has flat, broad leaves and lacks stiff spines.

Common saltwort, *Salsola kali* L., a native to sea beaches of Eastern Canada and the Old World, has shorter, broader leaves and longer bracts than Russian-thistle and is not weedy.

Halogeton glomeratus, a poisonous plant, is spreading rapidly in the Western United States. It is not known to occur in Canada (1968), but is mentioned here as it has been confused with Russian-thistle and might pass unnoticed. Halogeton has fleshy leaves, less than 2.5 cm long and round in cross section, with rounded tips ending in a bristlelike hair, not in a rigid spine as in Russian-thistle.

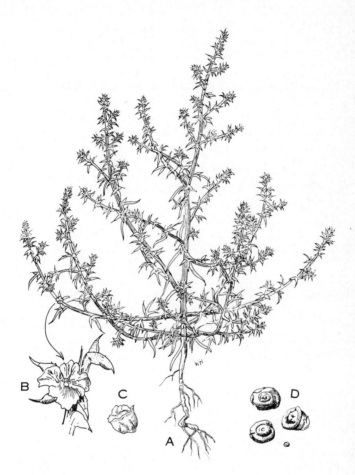

Russian-thistle, **Salsola pestifer: A,** plant; **B,** flower; **C,** seed covered by sepal bases; **D,** seeds.

AMARANTH FAMILY — AMARANTHACEAE

REDROOT PIGWEED *Amaranthus retroflexus* L.

Other names   Green amaranth, redroot, rough pigweed.

Description   A coarse annual, spreading by seed; taproot often pinkish or reddish; stems erect, usually 6-9 dm high and sometimes much taller, often branched, hairy at least above; leaves alternate, long-stalked, sparsely hairy, dull green, lower surface with prominent white veins; flowers numerous, small, green, crowded into dense fingerlike spikes forming a long terminal panicle, and in the leaf axils below, each flower enclosed in 3 bracts, the stiff, awl-shaped bracts longer than the flower and responsible for the bristly appearance of the spikes; seeds 1 mm broad, jet black, and glossy when mature, reddish at earlier stages, somewhat flattened. Flowering from July to September; maturing seed from August to October.

Origin   Considered to be native to North America, although its origin and history of spread are obscure. At present it is an abundant weed throughout the United States and Canada and is widely distributed elsewhere: Europe; Asia including Siberia, China, and Japan; New South Wales; and New Zealand.

Distribution in Canada   Redroot pigweed is present in the agricultural areas of all provinces except Newfoundland.

Habitats   Cultivated fields, gardens, roadsides, and waste places. Usually on rich soils.

Similar plants   Tumble pigweed, *Amaranthus albus* L., and prostrate pigweed or prostrate amaranth, *Amaranthus blitoides* S. Wats., are widely spread weeds that differ from redroot pigweed in having flowers in very small clusters at the leaf bases, rather than in conspicuous spikes. Tumble pigweed is an erect plant, with long spiny-tipped bracts and seeds somewhat smaller than in redroot pigweed. Prostrate pigweed forms mats and has very short bracts and larger seeds about 1.5 mm broad. Some of these differences are shown in the plate on the opposite page.

Green pigweed, *Amaranthus powellii* S. Wats., apparently native to western North America, has migrated eastward in recent times. It occurs in British Columbia and southwestern Ontario, where it is particularly abundant in orchards south of Hamilton. Green pigweed differs in many ways from redroot pigweed, but it is a more slender plant, the terminal spike is longer, the lateral spikes are less crowded, and the upper stem lacks the thick coating of hairs found on redroot pigweed.

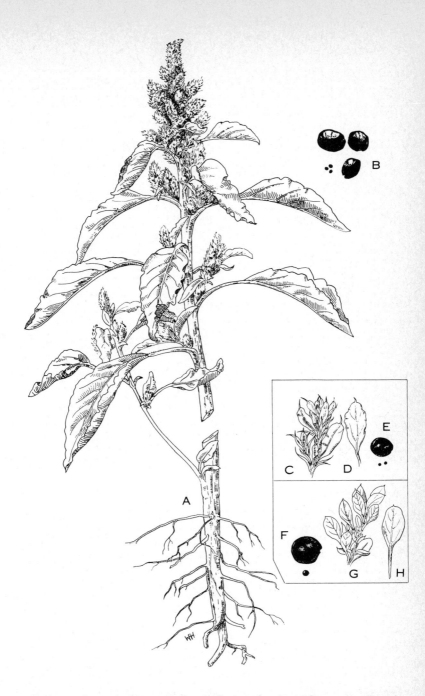

Redroot pigweed, **Amaranthus retroflexus: A,** plant; **B,** seeds. Tumble pigweed, **Amaranthus albus: C,** tip of branch showing leaves and long bracts; **D,** leaf; **E,** seed. Prostrate pigweed, **Amaranthus blitoides: F,** seed; **G,** tip of branch showing leaves and short bracts; **H,** leaf.

PURSLANE FAMILY — PORTULACACEAE

PURSLANE *Portulaca oleracea* L.

Other names   Pursley, pusley, pussley, wild portulaca.

Description   Annual; stems prostrate, stout and fleshy, often reddish, abundantly branched from the base and forming large circular mats; leaves clearly alternate below, grouped at the branch tips, thick and fleshy, hairless, rounded or flattened at the ends, and narrowed to the base; flowers stemless, in leaf axils or at stem tips, with 5 small yellow petals, petals soon falling, flowers opening only during the mornings of sunny days; seed capsule small and inconspicuous and resembling leaf buds, opening below the middle by a lid, containing many seeds; seeds very small, about 0.7 mm broad, flattened, rounded, or somewhat kidney-shaped, black and shiny, warty, with a whitish scar at one end. Flowering about the middle of July; maturing seeds shortly afterwards.

Origin   Europe. Probably Asia originally, now a cosmopolitan weed.

Distribution in Canada   Widely distributed. Occurs in every province except possibly Newfoundland.

Habitats   Particularly in gardens on rich soil but also in cultivated fields, waste places, and driveways.

Notes   Purslane has certain peculiarities that make it a troublesome weed: later germination than other weeds; long continued seed production; inconspicuous flowering and seeding; seed dormancy and retention of viability for many years; and the capacity of uprooted plants to ripen seed. Young shoots of purslane are often eaten as a cooked vegetable.

Similar plants   Purslane with its prostrate habit, succulent stems and leaves, and yellow flowers is not easily confused with other plants. The prostrate annual spurges (*Euphorbia* spp.) are often pests in gardens. They are not succulent, have smaller greenish flowers, and when their stems and leaves are broken, a milky juice lacking in purslane is visible.

Purslane, **Portulaca oleracea: A,** plant; **B,** seeds.

PINK FAMILY — CARYOPHYLLACEAE

CORN SPURRY *Spergula arvensis* L.

Other names   Devil's-gut, pickpurse, sandweed.

Description   Annual; several stems from the base, 15-45 cm high, sparsely hairy and somewhat sticky, slender; leaves in clusters, needlelike, rounded on the upper surface and grooved lengthwise on the lower surface, about 2.5 cm long, stipules small, thin, and yellowish brown; flowers about 6 mm across, numerous, in leafless terminal forked inflorescences, on slender stalks about 12 mm long, the slender stalks turn abruptly downwards as the capsule matures and later become upright again; petals 5, white, shorter than or equaling the sepals; sepals 5 and separate except at the base, styles 5; seed capsule nearly twice as long as the sepals, splitting into 5 entire parts, many seeded; seeds with a narrow light-colored wing, usually about 1 mm across but sometimes as large as 1.5 mm across, blackish, circular except for a small notch at one side, rounded on both surfaces, surfaces covered with light-colored club-shaped protuberances, or without these protuberances in the rare var. *sativa* (Boenn.) Mert. & Koch. Flowering from June to October; maturing seed from July to October.

Origin   Probably Europe. Now almost cosmopolitan.

Distribution in Canada   Known to occur in all provinces except Manitoba and Saskatchewan. Common only from Quebec eastwards and in southwestern British Columbia.

Habitats   Grainfields, cultivated fields, gardens, and roadsides. Considered to be an indicator of acid soils in Europe.

Similar plants   The family to which corn spurry belongs contains many weedy species. These species are usually hard to identify unless careful attention is given to the styles, sepals, capsules, and seeds. Corn spurry is rather closely marked by its clusters of leaves, thin dry stipules, and circular narrowly winged seeds. It might be confused with the sand spurries, *Spergularia* spp., which also have stipules but differ in having opposite leaves, capsules splitting into three not five parts, seeds somewhat kidney-shaped and certainly not round, seeds sometimes broadly winged.

A variety of corn spurry, var. *sativa*, sometimes considered to be a separate species, *Spergula sativa* Boenn., is mentioned in the description. It differs in having seeds without minute club-shaped outgrowths; otherwise it is similar, although it has a tendency to be more glandular hairy, whence the common name, sticky spurry. This variety is known only in Quebec, Alberta, and British Columbia.

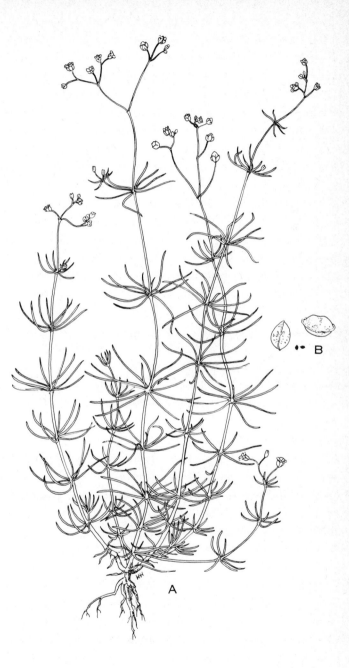

Corn spurry, **Spergula arvensis: A,** plant; **B,** seeds.

PINK FAMILY — CARYOPHYLLACEAE

CHICKWEED *Stellaria media* (L.) Vill.

Description   Annual or winter annual; stems usually prostrate, sometimes ascending, branching, round in cross section, with a conspicuous line of hairs on one side, rooting at the nodes and forming mats; leaves very variable in size in different plants, opposite, broadly oval and pointed, entire, the lower leaves stalked and the upper sessile, stalks of the lower leaves with a line of hairs, otherwise leaves glabrous; flowers about 6 mm across, star-shaped, single in the leaf axils, short-stalked, stalk elongating as the seeds mature; petals 5, white, shorter than the sepals; sepals 5, separate, hairy; styles 3; seed capsules many-seeded, longer than the sepals, opening by 6 teeth; seeds yellowish to dark brown, barely 1 mm across, almost circular but somewhat narrowed to the base, flattened, roughened, each side covered by 5 or 6 curved rows of small, coarse, not definitely elongate, protuberances. Flowering throughout the growing season, maturing seed rapidly after flowering.

Origin   Europe. Now a cosmopolitan weed.

Distribution in Canada   In the agricultural areas of all provinces, although apparently more common in British Columbia and Eastern Canada than in the Prairie Provinces.

Habitats   Grainfields, cultivated fields, pastures, gardens, lawns, and waste places.

Similar plants   A useful character for identification of chickweed is the single line of hairs down one side of the stem and on the leaf stalks. The broad leaves of chickweed are quite unlike those of other chickweeds.

Several other species related to chickweed are found in Canada. The weediest of these plants is the grass-leaved stitchwort, *Stellaria graminea* L., originating from Europe. This species probably occurs in every province, although rare in Western Canada, and is at least as abundant as chickweed in the Maritimes. Its square stems, narrow and stalkless leaves, petals longer than the sepals, and the elongated wavy protuberances on the seeds permit easy distinction from chickweed.

See the description of the mouse-ear chickweed, *Cerastium vulgatum* L., on page 62.

Chickweed, **Stellaria media: A,** plant; **B,** seed. Grass-leaved stitchwort, **Stellaria graminea: C,** plant; **D,** seed.

PINK FAMILY — CARYOPHYLLACEAE

MOUSE-EAR CHICKWEED *Cerastium vulgatum* L.

Other names   The scientific name used for this plant in the most recent European flora is *Cerastium fontanum* Baumg. ssp. *triviale* (Link) Jalas.

Description   Perennial, forming patches; stems spreading and sometimes rooting at the nodes, the flowering stems becoming erect, hairy, and sometimes glandular as in forma *glandulosum* (Boenn.) Druce; leaves opposite, usually about 12 mm long, rather long-hairy, oblong, sessile; flowers at the tips of branches, in groups that at first are compact and later become rather open; petals 5, white, deeply divided, about as long as the sepals; sepals hairy, with whitish margins; styles 5; seed capsules pale in color, narrow and up to twice as long as the sepals, 8 mm long, opening by 10 short teeth, many seeded; seeds small, about 0.8 mm long, slightly flattened, 4-sided, usually narrowed to the base but sometimes nearly circular, surfaces covered with elongated protuberances. Flowering throughout the growing season.

Origin   Europe.

Distribution in Canada   In every province; most abundant on the Pacific Coast and east of the Great Lakes.

Habitats   Lawns, pastures, and cultivated land.

Similar plants   The hairy stalkless leaves, small flowers, and prostrate habit of mouse-ear chickweed are sufficiently characteristic for identification.

A closely related plant, field chickweed, *Cerastium arvense* L., is found throughout Canada in meadows, pastures, open prairies, and other habitats. Field chickweed is native to Europe and North America. It is perennial and, like mouse-ear chickweed, forms patches. It is easily distinguished from this species by its showy flowers that have petals two or three times as long as the sepals, and by its narrower leaves.

Although *Stellaria media* and its close allies (page 60) are similar to the chickweeds described here, some fairly obvious differences exist. The chickweeds of the *Stellaria* group are less hairy, their flowers have only 3 styles, and the capsules containing the seeds open completely into 6 parts, not as in the mouse-ear chickweed by 10 slender teeth.

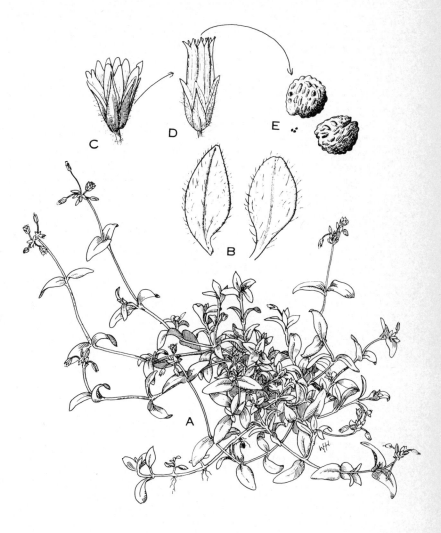

Mouse-ear chickweed, **Cerastium vulgatum: A,** plant; **B,** upper and lower leaf surfaces showing hairs; **C,** flower; **D,** seed capsule; **E,** seeds.

PINK FAMILY — CARYOPHYLLACEAE

WHITE COCKLE *Lychnis alba* Mill.

Other names   Evening lychnis, white campion.   The scientific name *Melandrium album* (Mill.) Garcke previously used for this plant in Europe has been superseded by *Silene alba* (Mill.) E. H. L. Krause.

Description   Biennial or short-lived perennial with thick fleshy roots; stems to 9 dm high, hairy; leaves opposite, pointed, entire, hairy; flowers large, fragrant, petals white or rarely pink and longer than the sepals, sepals (calyx) hairy and united into a tube; flowers either male or female, the sexes on different plants; male flowers with 10 stamens, calyx with 10 veins; female flowers with 5 styles and a seed capsule opening by 10 teeth, calyx with 20 veins; seeds kidney-shaped, grayish, about 1.3 mm broad, closely warted. Flowering June to September.

Origin   Europe.

Distribution in Canada   Found in all provinces, but less common in the Prairie Provinces.

Habitats   Hayfields, grainfields, roadsides, railway tracks, and waste places.

Similar plants   White cockle is often mistaken for night-flowering catchfly or sticky cockle, *Silene noctiflora* L., originating from Europe. This catchfly is so closely related to white cockle that in Eurasia botanists place both plants in the same genus. Night-flowering catchfly is found in all provinces and is often more common than white cockle.

These superficially similar plants can be readily differentiated in life by squeezing the plant parts between the fingers. Night-flowering catchfly is distinctly sticky, while white cockle is not. There are some other well-marked differences: the catchfly is annual or winter annual, both sexes are in the same flower, its capsules are surmounted by 3 styles and open by 6 teeth, the calyx teeth are distinctly long-tapered and the 10 green veins on the calyx are branched (see the illustration on the opposite page).

The red campion, *Lychnis dioica* L., similar to white cockle but with red petals, is mentioned in North American weed manuals. Red campion is native to Europe where it is found in open woods but not as a weed of fields. European botanists consider the cockles with pink petals occurring in cultivated land to be hybrids of the red campion and the white cockle. The pink-petalled plants very rarely found as field weeds in Canada are probably not red campion but hybrids that, apart from flower color, are inseparable from white cockle.

White cockle, **Lychnis alba: A**, plant; **B**, calyx; **C**, seed. Night-flowering catchfly, **Silene noctiflora: D**, part of flowering stem showing young buds and an open flower; **E**, calyx (note strong veins); **F**, seed.

PINK FAMILY — CARYOPHYLLACEAE

BLADDER CAMPION *Silene cucubalus* Wibel

Other names   Cow-bell, rattleweed. Several scientific names have been used for this plant other than the one above: *Silene vulgaris* (Moench) Garcke; *Silene latifolia* (Mill.) Britten & Rendle; *Silene inflata* Sm.

Description   Perennial by deep persistent roots, spreading by seeds and parts of the crown severed by implements; stems several from the base and forming large clumps, usually about 45 cm high, smooth; leaves opposite, without stalks, usually glabrous, very variable in size; flowers in a loose terminal cluster, styles 3, petals white and nearly twice as long as the sepals; sepals (calyx) united into an inflated, bladderlike, and veiny tube with 5 short triangular teeth, glabrous except for the teeth, calyx about 12 mm long; capsule enclosed within the calyx, almost spherical, opening by 6 small teeth; seeds kidney-shaped, closely warted, 1.5 mm across, grayish. Flowering from mid-June to September and sometimes later.

Origin   Eurasia.

Distribution in Canada   Widely distributed and occurring in every province. More common east than west of the Great Lakes.

Habitats   Hayfields, cultivated fields, waste places, and roadsides.

Similar plants   Night-flowering catchfly and white cockle are sometimes mistaken for bladder campion. Bladder campion is practically glabrous and its calyx teeth are short, broad, and triangular, whereas the catchfly and cockle are quite hairy and the teeth of the calyx elongate.

In recent years, biennial campion, *Silene cserei* Baumg., Asiatic in origin, has become established in Canada. Superficially similar to bladder campion it differs in the following ways: rosette-forming, biennial or possibly short-lived perennial; leaves usually thicker and broader; calyx smaller and little inflated at maturity and with obscure net veins, calyx opening constricted; seed capsules egg-shaped and longer than the calyx (nearly spherical and hidden in the calyx in bladder campion); seeds smaller. The most precise character for differentiation is a technical one requiring magnification: the short capsule stalk (carpophore) above the calyx is minutely hairy in biennial campion and glabrous in bladder campion. Biennial campion is found in all the western provinces, Ontario, and Quebec. Although mostly a pest around railway installations, it may become troublesome in abandoned fields along railway lines.

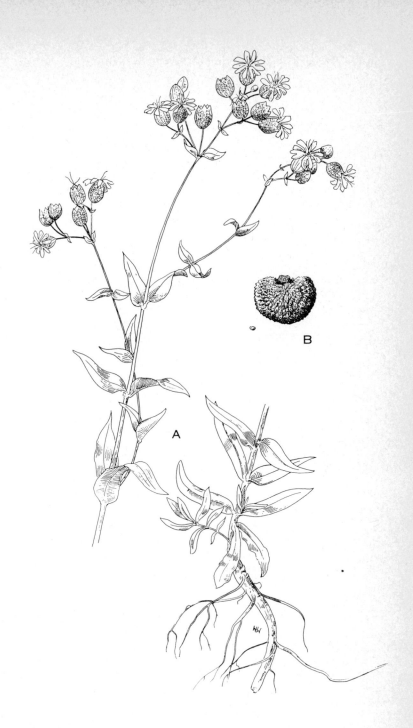

Bladder campion, **Silene cucubalus: A,** plant; **B,** seed.

PINK FAMILY — CARYOPHYLLACEAE

COW COCKLE *Saponaria vaccaria* L.

Other names   China cockle, cow-herb. *Vaccaria pyramidata* Med. *Vaccaria segetalis* (Neck.) Garcke, and *Vaccaria vulgaris* Host are scientific names used for this species in Europe. These names are rarely used here.

Description   Annual; stems branched, glabrous, whitish, 15-60 cm high, swollen at nodes; leaves opposite, all sessile and glabrous, clasping the stem, acute at the tip, oblong with the upper leaves often broader towards the base, thick and fleshy; flowers showy, loosely grouped at the ends of the stems, styles 2, petals pale red or rarely white, exceeding the sepals; sepals (calyx) united in a tube that is about 12 mm long, inflated, flask-shaped, strongly 5-ribbed, the ribs becoming broad, dark green flanges at maturity; seed capsules spherical, opening by 4 teeth, many seeded; seeds nearly 2.5 mm across, roundish, orange to dull black, the surface roughened by numerous minute points. Flowering from June to September.

Origin   Eurasia.

Distribution in Canada   Troublesome only in the Prairie Provinces. More common in Saskatchewan than elsewhere. Absent only in Prince Edward Island and Newfoundland.

Habitats   Grainfields in Western Canada, waste places, roadsides, and railway grades. Most troublesome in fine-textured soils.

Notes   The seeds contain saponin and they are poisonous to animals.

Similar plants   Cow cockle is easily recognized by its pale red flowers, opposite and somewhat fleshy leaves, lack of hairs on all parts, and the broad, green flanges on its otherwise light-colored calyx. Purple cockle or corn cockle, *Agrostemma githago* L., a Eurasiatic plant, is an annual with purplish flowers, and a prominently ribbed calyx. It differs from cow cockle in being very hairy and in the calyx teeth being conspicuous and exceeding the petals in length. Purple cockle is now rare as a weed.

Bouncingbet or soapwort, *Saponaria officinalis* L., introduced from Europe as a garden plant, is a common roadside weed in southern Ontario and southern Quebec. It is rare in the western provinces and has only recently been recognized as part of the weed flora of Manitoba and Newfoundland. It can be easily differentiated from cow cockle by its perennial habit, white or light pink flowers in dense terminal clusters, and its unwinged calyx.

Cow cockle, **Saponaria vaccaria: A,** plant; **B,** calyx of mature flower; **C,** seed.

BUTTERCUP FAMILY — RANUNCULACEAE

TALL BUTTERCUP *Ranunculus acris* L.

Other names   Meadow buttercup, tall crowfoot, tall field buttercup.

Description   Perennial with very short rootstock and thick roots, spreading by seeds; stems erect, 4-9 dm high, clustered, branched, hairy; leaves alternate, soft hairy on both surfaces, deeply divided almost to the base, only the less divided upper leaves below the flowers without stalks; flowers about 2.5 cm across, grouped on long stalks at the top of the plant; petals 5, bright yellow, longer than the sepals; stamens numerous; seeds in rounded groups, numerous, each seed 3 mm long, strongly flattened, tipped by a short hooked beak. Flowering from May to September.

Origin   Europe.

Distribution in Canada   Tall buttercup is one of the most abundant weeds of the six eastern provinces. Although far less common in Western Canada, it appears to be increasing in the following places: in the areas south of Lake Manitoba and north of Swan River, in Manitoba; in the irrigated areas of southern Alberta; and in southwestern British Columbia, particularly in pastures of the Fraser Valley.

Habitats   Pastures, meadows, stream and irrigation ditch banks, and roadsides; particularly troublesome in permanent pastures on poorly drained soil. Tall buttercup, usually shunned by cattle, produces seeds freely and forms conspicuous clumps on otherwise well-grazed pastures.

Notes   Buttercups contain an acrid and bitter juice and, when grazed, cause severe pain and serious inflammation. Buttercups in hay are considered to be harmless, as the poisonous principle is volatile and is dispelled when the hay is cured.

Similar plants   There are many native and a few introduced species of buttercup in Canada. Only tall buttercup is a weed of primary importance. Creeping buttercup, *Ranunculus repens* L., is a European weed, widely distributed in Canada in lawns and waste places. Although usually a plant of creeping habit, it does exist as an erect plant without runners (var. *erectus* DC.) and is then difficult to separate from tall buttercup unless attention is directed to the leaf divisions: distinctly stalked in creeping buttercup, not stalked in tall buttercup.

Tall buttercup, **Ranunculus acris: A,** plant; **B,** seed.

BARBERRY FAMILY — BERBERIDACEAE

COMMON BARBERRY *Berberis vulgaris* L.

Description   A tall, erect shrub, usually 12-24 dm high, many stemmed from the base, bark light gray, but yellowish on first-year branches, branches often arching, wood yellow; leaves 18-50 mm long, bright green or purplish in var. *atropurpurea* Rgl., oval, edges finely spiny-toothed, short-stalked, alternate or clustered in short shoots with spines at the base of the shoots, spines usually 3-branched; flowers in loose drooping clusters, yellow, 6 mm across; berries oblong and bright red, at least 8 mm long, sour, 1 to several seeded; seeds nearly 6 mm long, oblong, chocolate brown, shiny, wrinkled. Flowering from late May to early June, berries maturing by September and remaining on the shrubs through the winter.

Origin   Europe.

Distribution in Canada   The largest stands of common barberry occur in southern Ontario and Quebec.

Notes   Stem rust is a serious fungus disease of wheat, oats, barley, and grasses. Wherever barberry is present, severe local epidemics of rust may occur. The black spores of the rust overwinter on stubble and infect barberry. Barberry bushes permit early infections of red spores on nearby grains. In Western Canada, infections of stem rust are mainly caused by showers of red spores from the south.

Common barberry and hybrids between it and Japanese barberry, *Berberis thunbergii* DC., have largely been eradicated from the Prairie Provinces. A vigorous campaign to eradicate these plants from Ontario and Quebec was started in 1964 by the federal and provincial governments. The importation, propagation, and interprovincial movement of all deciduous barberry, including Japanese barberry, is now prohibited.

Similar plants   Japanese barberry is a shorter shrub than common barberry, and has smaller leaves, smooth leaf edges, usually unbranched spines, and flowers and berries solitary or in clusters of two to four.

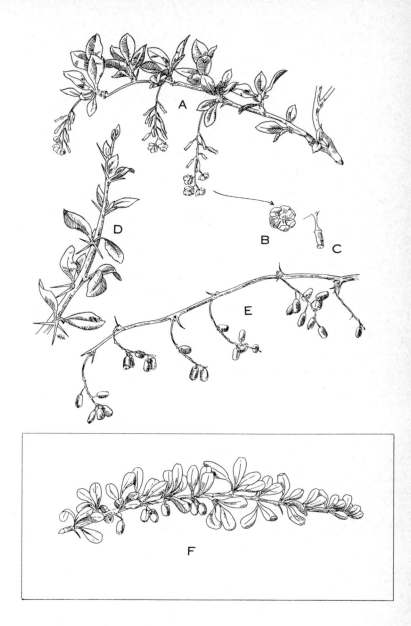

Common barberry, **Berberis vulgaris: A,** branch with flowers; **B,** flower; **C,** maturing berry; **D,** young shoot with branched spines; **E,** branch and mature berries. Japanese barberry, **Berberis thunbergii: F,** branch with berries.

MUSTARD FAMILY *Cruciferae*

The mustard family is of much economic significance as it contains not only many troublesome weeds but also important food plants such as cabbage, rutabaga, turnip, cauliflower, and radish, and ornamentals such as rocket, stock, honesty, and wallflower. Plants belonging to this family are among our worst weeds because they seed prolifically, grow rapidly, and are adapted to a wide range of environmental conditions.

The family is characterized by having flowers with 4 separate sepals and 4 separate petals arranged in 2 opposite pairs, and so forming a cross-shaped flower. There are 6 stamens, 2 shorter than the others. The flowers on stalks are arranged at the tips of stems. The seeds are borne in pods that usually consist of 2 outside walls separated by a thin white partition. The pods in many species open from below, as shown in the illustration (*D*). In other species such as ball mustard, the pods are indehiscent, that is, they do not open at maturity to release seeds. In the wild radish, the pods break across into 1-seeded joints. Star-shaped and branched hairs are common in this family and are useful in identification, particularly before pods are formed.

Although members of this family are readily recognized, individual species are hard to identify. It is helpful and sometimes essential to examine mature plants so that the following points may be observed: length and angle of the short stalk bearing the pod, presence or absence of a beak, length of beak, breadth and length of pods, and characters of the seeds.

A number of pods are shown on the opposite page (*E* to *H*). None of these species are illustrated elsewhere and brief descriptions are given below:

Hoary alyssum, *Berteroa incana* (L.) DC. (*E*). Perennial. White flowers. Pods with star-shaped hairs. From Europe. Occurs in British Columbia, Saskatchewan, Manitoba, Ontario, Quebec, New Brunswick, and Nova Scotia.

Creeping yellow cress, *Rorippa sylvestris* (L.) Bess. (*F*). Perennial, spreading by underground stems. Yellow flowers. From Europe. Occurs in all provinces.

Marsh yellow cress, *Rorippa islandica* (Oeder) Borbas, (*G*) and varieties. Annual or biennial. Yellow flowers. Native to Canada. Occurs in all provinces, usually in wettish sites.

Wood whitlow-grass, *Draba nemorosa* L. (*H*). Annual or winter annual. Pale yellow flowers. Native to Canada. Occurs in western Ontario and the western provinces.

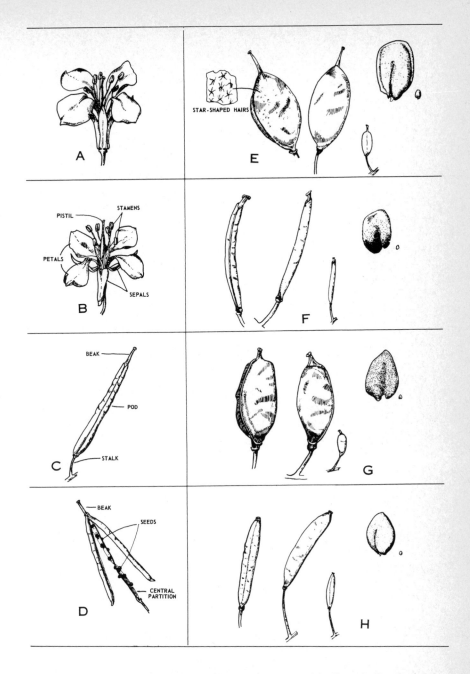

A, Mustard flower; **B**, mustard flower (diagrammatic); **C**, pod; **D**, opening pod showing seeds on central partition. Enlarged pods, pod of normal size, and seeds of the following species: **E**, hoary alyssum, **Berteroa incana**; **F**, creeping yellow cress, **Rorippa sylvestris**; **G**, marsh yellow cress, **Rorippa islandica**; **H**, wood-whitlow grass, **Draba nemorosa.**

STINKWEED *Thlaspi arvense* L.

Other names   Fanweed, pennycress.

Description   Annual or winter annual with an unpleasant odor when bruised, spreading by seeds, entire plant glabrous; stems erect, 5 cm to 6 dm high, simple or branched above; leaves alternate, glabrous, not divided, leaf edges distantly and irregularly toothed; basal leaves with stalks, soon withering; stem leaves clasping the stem by 2 earlike lobes; flowers about 3 mm across, petals white; pods on slender upwardly curving stalks; pods bright green, becoming yellowish or greenish orange, almost circular, 12 mm across, strongly flattened and winged, wings with a deep narrow notch at the top, each pod containing from 4 to 16 seeds; seeds 2 mm long, reddish brown to black, somewhat flattened, with several rows of curved ridges on each side. Flowering from early spring to late autumn with the peak in June.

Origin   Europe and Asia. Introduced to North America at a very early date. According to an early writer, Thomas Nuttall, it was a common weed around Detroit by 1818.

Distribution in Canada   Occurs in all provinces, but is most abundant and troublesome on the prairies west from the heavy soils of the Red River valley in Manitoba.

Habitats   Grainfields, hayfields, gardens, and waste places. A serious pest in the areas of intensive grain growing in Western Canada. In Eastern Canada, mainly a weed of waste places, although occasionally a crop weed.

Notes   Before flowering, stinkweed can be readily identified by the rank smell when its leaves are crushed.

When grazed, stinkweed produces a taint in dairy products. Feeds containing excessive amounts of stinkweed seed may be poisonous to horses, cattle, and pigs, and may produce off-flavors in meat products.

Similar plants   The pepper-grasses resemble stinkweed in having white flowers, and flattened and notched pods. The flowers and pods of the pepper-grasses are much smaller than those of stinkweed, and the pods contain only two seeds.

Stinkweed, **Thlaspi arvense: A,** plant; **B,** upper part of plant with maturing
pods; **C,** mature pods; **D,** opening pod and seed.

MUSTARD FAMILY — CRUCIFERAE

COMMON PEPPER-GRASS *Lepidium densiflorum* Schrad.

Other names   Green-flowered pepper-grass.

Description   Annual or winter annual, very bushy, spreading only by seeds; stems 20 cm to 6 dm high, freely branched above, covered with short simple hairs; leaves alternate, distantly toothed or deeply divided, not clasping; flowers small, petals absent or shorter than the sepals, white; pods nearly round, about 2 mm broad, with a narrow notch at the top, strongly flattened, on stalks about as long as the pods, each half of the pod containing a single seed; seeds oblong, flattened, about 1.5 mm long, bright reddish yellow. Flowering June to August.

Origin   Native to North America, but now introduced in many parts of the world.

Distribution in Canada   A very common weed occurring in all provinces.

Habitats   Fields, waste places, roadsides, and gardens.

Similar plants   Eleven species of pepper-grasses occur in Canada. Some of these species are hard to distinguish from the common pepper-grass. Poor man's pepper-grass, *Lepidium virginicum* L., also a native plant of wide distribution, has petals about twice as long as the sepals and this is the most practical characteristic for differentiation from the common pepper-grass, which either lacks petals or has very reduced petals shorter than the sepals.

Field pepper-grass, *Lepidium campestre* (L.) R. Br., an introduction from Europe, is rather abundant in some parts of Eastern Canada and British Columbia, and is known to occur in all provinces except Manitoba. As the illustration shows, the upper leaves clasp the stem and the tip of the pod is curved upward and broadly winged. Also, the pods are covered with small bladderlike scales. This combination of characters separates this pepper-grass from other species.

Clasping-leaved pepper-grass, *Lepidium perfoliatum* L., introduced from Europe, has been found in Quebec but is as yet only of importance in drier regions of the west, where it occurs in southwestern Saskatchewan, southern Alberta, and the inland valleys of British Columbia. It is easily distinguished from other pepper-grasses. Particularly characteristic are the pale yellow flowers, finely dissected lower leaves, and the rounded upper leaves completely encircling the stem.

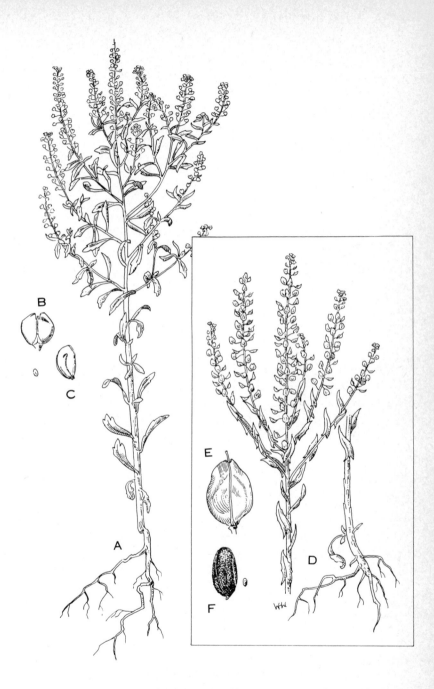

Common pepper-grass, **Lepidium densiflorum: A,** plant; **B,** pod; **C,** seed.
Field pepper-grass, **Lepidium campestre: D,** plant; **E,** pod; **F,** seed.

HEART-PODDED HOARY CRESS *Cardaria draba* (L.) Desv.

Other names   Perennial pepper-grass, white-top, white-weed. Formerly known by the scientific name *Lepidium draba* L.

Description   Perennial, forming dense patches, spreading by large white root-stocks and seeds; stems 20 cm to 6 dm high, branched, often sprawling; leaves 12-75 mm long, alternate, broad, oblong, irregularly toothed to almost entire, with short unbranched hairs on both surfaces; upper leaves clasping the stem with earlike lobes; flowers numerous in a flat-topped cluster, showy, small, about 3 mm across, petals white; pods glabrous, heart-shaped at base, flattened, about 2.5 mm long, topped by a short persistent style, pods on stalks up to 12 mm long, each half of the pod with 1 seed; seeds about 2 mm long, oval, slightly flattened, reddish brown. Flowering May to July.

Origin   This weed may have been introduced into Canada from two sources: Europe and Western Asia. It was first collected in Canada at Barrie, Ontario, in 1878.

Distribution in Canada   Heart-podded hoary cress is the most widely distributed hoary cress in Canada. It occurs in all provinces except Prince Edward Island, New Brunswick, and Newfoundland.

Habitats   Grainfields, hayfields, and roadsides. Often abundant on alkali soil.

Similar plants   Two other hoary cresses are now known to occur in Canada. In Western Canada, where these three hoary cresses occur, heart-podded is the least abundant. These species are superficially similar, but the seedpods afford striking differences recognizable even before maturity.

Lens-podded hoary cress, *Cardaria chalepensis* (L.) Handel-Mazzetti, which occurs in the Prairie Provinces, British Columbia, and Ontario, was almost certainly introduced into Western Canada with Turkestan alfalfa. Pods of this hoary cress are nearly round in outline and contain two to four seeds.

Globe-podded hoary cress, *Cardaria pubescens* (Meyer) Jarmolenko, often occurs with lens-podded and has the same general distribution and origin. It differs from the other two hoary cresses in usually having smaller and narrower leaves but, above all, in its minutely hairy, globular and inflated pods. The pods contain two to four seeds.

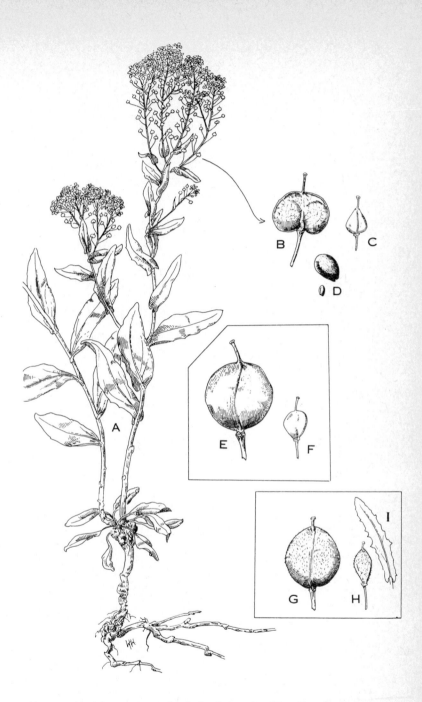

Heart-podded hoary cress, **Cardaria draba: A,** plant; **B,** pod; **C,** immature
pod; **D,** seed. Lens-podded hoary cress, **Cardaria chalepensis: E,** pod;
F, immature pod. Globe-podded hoary cress, **Cardaria pubescens:
G,** pod; **H,** immature pod; **I,** leaf.

MUSTARD FAMILY — CRUCIFERAE

SHEPHERD'S PURSE *Capsella bursa-pastoris* (L.) Medic.

Description   Annual or winter annual, spreading only by seeds; stems solitary or several from the same root, simple or branched, 3 cm to 9 dm high; leaves usually with scattered simple and branched hairs; basal leaves, when present, in a rosette, narrowed into a stalk, very variable, usually deeply cleft but sometimes entire; stem leaves also variable in outline but always stalkless, alternate, clasping the stem with earlike projections; flowers small, about 2.5 mm wide, petals white and exceeding the sepals; pods about 6 mm long, flattened, triangular, notched at the top, pods on spreading stalks 5-20 mm long, each pod containing about 20 seeds; seeds 1 mm long, oblong, orange-yellow, surface dull and punctured. Flowering from early spring to late fall.

Origin   A European plant introduced into North America before 1700. One of the most widely distributed weeds occurring in wasteland and cultivated areas everywhere except in the tropics.

Distribution in Canada   One of our most common weeds, found in the agricultural areas of all provinces.

Habitats   Grainfields, cultivated fields, paths, gardens, and roadsides.

Notes   Shepherd's purse commonly harbors fungi, which are transmitted to cabbage, turnips, and other members of the mustard family.

Similar plants   Although shepherd's purse is variable in height and leaf shape, the pods are reasonably constant in outline and quite distinctive even at early stages. Stinkweed has much larger pods and is completely glabrous, while shepherd's purse is usually hairy.

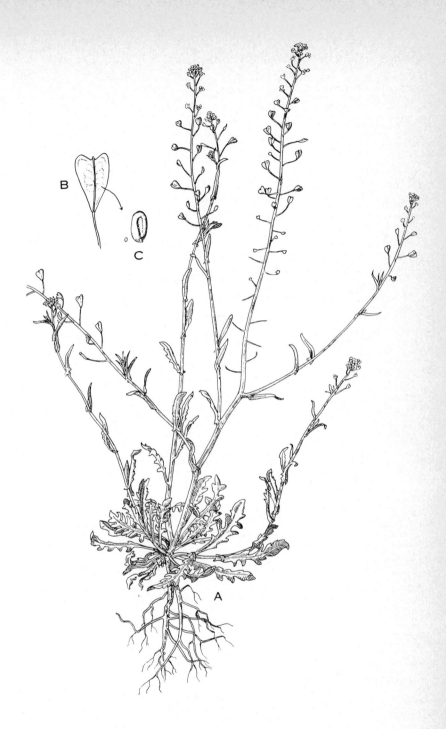

Shepherd's purse, **Capsella bursa-pastoris: A,** plant; **B,** pod; **C,** seed.

MUSTARD FAMILY — CRUCIFERAE

BALL MUSTARD *Neslia paniculata* (L.) Desv.

Other names   Neslia, yellow weed.

Description   Annual or winter annual with a slender taproot, spreading by seeds; stems erect, slender, 3-9 dm high, branched above, with small star-shaped hairs especially below; leaves entire or occasionally with a few blunt teeth, sparsely covered with small star-shaped hairs; stem leaves alternate, arrow-shaped, clasping the stem by pointed lobes; flowers clustered at the ends of the stems, about 3 mm across, petals bright yellow; pods borne on slender spreading stalks, stalks about 9 mm long; pods small, about 3 mm across, covered with net-veined ridges when mature, roundish but broader than high, firm walled, indehiscent, that is, not splitting at maturity; seeds yellowish orange, with a prominent ridge formed by the embryo root, usually only one seed in a pod. Flowering June to September.

Origin   Europe.

Distribution in Canada   All provinces, the District of Mackenzie, and the Yukon. Common in Western Canada, particularly in the Peace River district. Not very common in Eastern Canada; serious infestations were recently observed in the Lake St. John area of Quebec.

Ball mustard was collected as early as 1891 at Portage la Prairie, Manitoba.

Habitats   Grainfields, particularly in the west, railway lines, and waste places.

Notes   The pods of ball mustard are removed with the crop and, because of their size and the fact that they do not open when ripe, they are hard to clean from wheat. Very commonly the pod is called a seed or a seed ball.

Similar plants   Ball mustard is a distinctive plant. The combination of yellow flowers, clutching leaves with star-shaped hairs, and small roundish net-veined pods should be sufficient for differentiation from other plants in the mustard family.

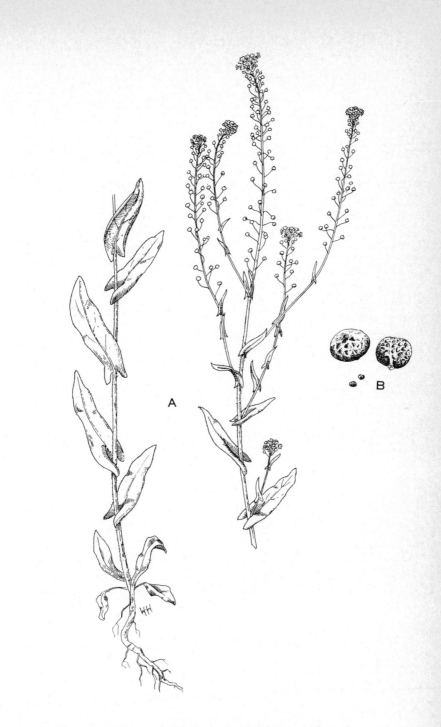

Ball mustard, **Neslia paniculata: A,** plant; **B,** pods.

SMALL-SEEDED FALSE FLAX *Camelina microcarpa* Andrz. ex DC.

Description   Annual or winter annual, spreading by seeds; stems 3-9 dm high, branched above, with long simple hairs and short star-shaped hairs particularly below or without long hairs; stem leaves alternate, stalkless, lanceolate or arrow-shaped, clasping; flowers pale yellow, small, in a terminal cluster; pods pear-shaped, strongly margined, inflated, less than 6 mm long, scattered along the elongated flowering stem on slender spreading stalks; seeds reddish brown, 1-1.3 mm long, oblong, warty, 10 or more seeds in each pod. Flowering from late May to August.

Origin   Europe and Asia.

Distribution in Canada   Occurs in all provinces, but much commoner in the west.

Habitats   In grain, flax, clover, alfalfa, on roadsides, along railways, and in waste places.

Similar plants Other false flaxes may be differentiated from this species by seed and pod characters. Large-seeded false flax, *Camelina sativa* (L.) Crantz, has less hairy stems, pods of the same shape over 6 mm long, and seeds about 2 mm long. It has about the same distribution as small-seeded false flax but is less abundant.

Flat-seeded false flax, *Camelina parodii* Ibarra & La Porte, has large seeds, about 2.5 mm long, and its seeds are flat and round, rather than oblong as in other species of false flax. Pods of this plant narrow to the base from a flat top and are top-shaped rather than pear-shaped. The scientific name *Camelina dentata* Pers. has been used for this flat-seeded plant in several Canadian publications.

Hoary alyssum, *Berteroa incana* (L.) DC. (page 75, Figure *E*), should not be confused with the false flaxes, although its pods are of rather similar size and shape. Hoary alyssum differs from the false flaxes in having: pods on erect stalks and touching or nearly touching the stem; pods covered with star-shaped hairs rather than being glabrous as in the false flaxes; leaves grayish from star-shaped hairs and not clasping; flowers white; seeds with a narrow wing. Hoary alyssum occurs sporadically across Canada and appears to be increasing, particularly in British Columbia.

Small-seeded false flax, **Camelina microcarpa: A,** plant; **B,** pod; **C,** seeds.
Large-seeded flax, **Camelina sativa: D,** seeds.

MUSTARD FAMILY — CRUCIFERAE

WILD RADISH *Raphanus raphanistrum* L.

Other names   Jointed charlock, jointed radish, jointed wild radish.

Description   Annual or winter annual with slender taproot; stems 3-9 dm high, freely branched, with coarse short hairs especially at the base; leaves alternate, to 7.5 cm long, usually rough; the lower leaves deeply divided and with a large terminal segment; uppermost leaves narrower and often entire; flowers 12-18 mm across; petals yellow, less commonly white or purple, conspicuously veined; pods 2.5-7.5 cm long, narrow, terminating in a long pointed beak, pods longitudinally ribbed, constricted between the seeds, when ripe splitting across at the joints into barrel-shaped fragments each of which contains a seed; seeds reddish brown, oval, 2.5 mm long, finely net veined.

Origin   Probably from Europe. Found in the whole of Europe, except the arctic regions; North Africa; Asia Minor; and Syria; rare in Siberia and the Far East. The native home of the wild radish may be in the Mediterranean region, but it is now known only as an agricultural weed. From Europe it has been introduced into South Africa, Japan, Australia, and North and South America.

Distribution in Canada   Wild radish is very abundant in all the provinces on the Atlantic seaboard and is apparently still spreading in that area. Inland in Quebec and Ontario this weed is of less importance, although it is known from a number of localities. Its distribution in the Prairie Provinces is not well understood but there are reports of occurrence in the moister parts of Saskatchewan. Locally abundant in southern Vancouver Island and in the Vancouver area.

Habitats   Grainfields, waste places, and roadsides.

Similar plants   Cultivated radish, *Raphanus sativus* L., occasionally persistent in gardens, is similar to wild radish in the aboveground parts, but the flowers are purplish, pink or white, never yellow, and the seedpods are spongy, lack distinct joints, and split in various ways at maturity, but not into units containing single seeds.

Wild radish has large flowers, lighter yellow than wild mustard with which it is sometimes confused. The constricted seedpods and veined petals of wild radish serve to differentiate the two plants. The pod of wild radish does not split lengthwise as in wild mustard and its relatives.

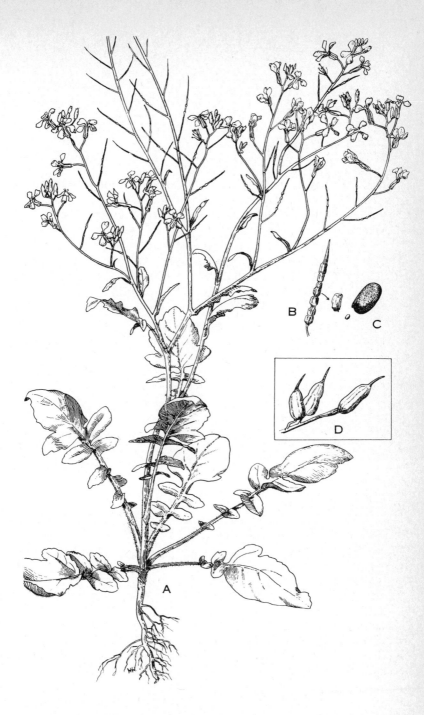

Wild radish, **Raphanus raphanistrum: A,** plant; **B,** pod; **C,** seed. Cultivated radish, **Raphanus sativus: D,** pods.

WILD MUSTARD *Sinapis arvensis* L.

Other names   Charlock. The scientific name *Brassica kaber* (DC.) L. C.
Wheeler var. *pinnatifida* (Stokes) L. C. Wheeler is often used for this plant by
North American botanists.

Description   Annual; stems 3-9 dm high, usually with stiff hairs at least at the
base, simple or branched, branches purple at their junction with the main stem;
leaves alternate, somewhat hairy, particularly on the veins of the lower surface;
lowest leaves stalked, deeply divided and consisting of a large terminal segment
and a few smaller lateral lobes; upper leaves stalkless, usually undivided but
coarsely toothed; flowers more than 12 mm across; petals bright yellow; pods
normally spreading from the stem on thick stalks that are less than 6 mm long;
pods glabrous or sometimes bristly hairy, usually prominently ribbed lengthwise, from 25-40 mm long without the beak; beak angular, about as wide as the
pod, 6-12 mm long, often containing a seed at the base; seeds round, about 1.5
mm across, black, appearing netted at high magnification. Flowering throughout the summer.

Origin   Eurasia.

Distribution in Canada   One of the commonest annual weeds. Occurs in all
provinces. The most serious infestations are probably in the rich river valleys of
the west.

Habitats   Grainfields, cultivated fields, waste places, fencerows, and roadsides.

Similar plants   Wild mustard and its close relatives (page 92) possess pods
with conspicuous beaks, rounded seeds, and flowers 12 mm or more across. The
groups of plants represented by flixweed (page 102) and tumble mustard (page
98) differ in having pods with inconspicuous beaks, small oblong seeds, and
small flowers usually much less than 12 mm across.

Some of the more significant characters separating wild mustard and related mustards (page 92) should be emphasized. In this group of five species,
only black mustard has pods appressed to the stem and less than 25 mm long,
white mustard is the only species with a flattened beak as long as or longer than
the body of the pod, and only bird rape has clutching leaves. The other two
species may be readily separated. Wild mustard is a dark green, somewhat hairy
plant with rather broad, toothed, upper leaves and its pods are on stout stalks
that are less than 6 mm long, whereas Indian mustard is a glabrous grayish plant
with narrow and entire upper leaves and its pods are on slender stalks about 12
mm long.

Wild mustard, **Sinapis arvensis: A,** plant; **B,** pod; **C,** seed.

INDIAN MUSTARD *Brassica juncea* (L.) Czern.

Description   Annual; stems 9-12 dm high; lower leaves deeply lobed, upper leaves narrow and entire; flowers about 12 mm across, petals pale yellow; pods spreading from the main stem, 25-50 mm long without the beak, beak 6-12 mm long, base of beak without a seed, pod stalks slender and nearly 12 mm long; seeds about 1.5 mm across, brownish red.

Origin and distribution in Canada   From Eurasia. Occurs in every province and reaches its greatest abundance in the western provinces.

BIRD RAPE *Brassica campestris* L.

Other name   Wild turnip.

Description   Annual or winter annual, a smooth bluish-green plant; stems 3-9 dm high; stem leaves entire, clasping the stem by earlike projections; flowers about 12 mm across, petals pale yellow; pods spreading from the main stem, 40-50 mm long without the beak, beak 6-12 mm long, base of beak seedless, pod stalks slender and from 12-25 mm or more in length; seeds about 1.5 mm across, nearly spherical, reddish brown or grayish black.

Origin and distribution in Canada   From Eurasia. Occurs in all provinces, sometimes abundant. In some parts of the east, bird rape supplants wild mustard over large areas.

BLACK MUSTARD *Brassica nigra* (L.) Koch

Description   Annual; stems 9-18 dm high, strongly branched; leaves all stalked, the terminal segment of the lower leaves large and coarsely toothed, the side lobes small, uppermost leaves narrow and entire; flowers about 12 mm across, petals bright yellow; pods erect and appressed to the stem, 12-18 mm long without the beak, beak about 3 mm long, pod stalks 3-6 mm long; seeds broadly oblong, dark brown.

Origin and distribution in Canada   From Eurasia. Not very common in Canada, although known from all provinces east of Alberta.

WHITE MUSTARD *Sinapis alba* L.

Description   Annual; stems 3-6 dm high; all leaves divided into irregular segments; flowers 12 mm across, petals pale yellow; pods bristly hairy, spreading widely from the main stem, 12 mm long without the beak, beak flattened and as long as the pod or longer, beak often with a seed at the base, pod stalks 8 mm long; seeds rounded, light yellow to light brown, about 3 mm across.

Origin and distribution in Canada   From Eurasia. Found in all provinces except those on the Atlantic Coast.

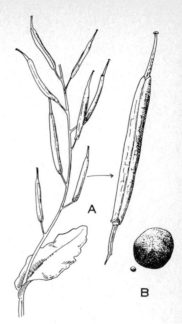

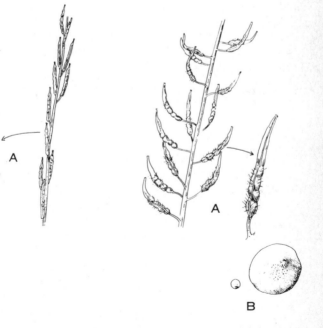

Mustards. **Upper left:** Indian mustard, **Brassica juncea. Upper right:** Bird rape, **Brassica campestris. Lower left:** Black mustard, **Brassica nigra. Lower right:** White mustard, **Sinapis alba.** In all drawings: **A,** group of pods and single pod; **B,** seed.

MUSTARD FAMILY — CRUCIFERAE

DOG MUSTARD *Erucastrum gallicum* (Willd.) O. E. Schulz

Description   Annual or winter annual, spreading by seeds; stems erect and branched, 15 cm to 6 dm high, at least the lower part of the stem with short white downward-directed hairs; leaves alternate, somewhat hairy, oblong in general outline, deeply cut into coarse oblong segments, not clasping the stem at the base; flowers 6 mm across, clustered at the top of the stem, petals pale yellow; pods 25-50 mm long, narrow, tipped with a short beak about 3 mm long, pods attached to the stem by slender stalks about 1/5 as long as the pod, many seeded; seeds in a single row in each half of the pod, 1.3 mm long, reddish brown, surface finely netted. Flowering in summer and late fall.

Origin   Europe. A comparatively recent introduction to North America. The first collection of this plant in the United States was made in 1903 and the first in Canada at Emerson, Manitoba, in 1922.

Distribution in Canada   Occurs in all provinces and reaches its greatest abundance in Manitoba and Saskatchewan.

Habitats   Roadsides, fields, waste places, along railways, gardens, and orchards. Very common on roadsides, an abundant field weed in many localities in Western Canada.

Similar plants   Dog mustard is the only plant of the mustard family in Canada with pods in the axils of leaves or leaflike bracts. By this distinctive character alone it should be readily distinguished from other mustards.

The leaves of dog mustard somewhat resemble the lower leaves of tumble mustard (page 98).

Dog mustard, **Erucastrum gallicum: A,** plant; **B,** flower; **C,** pod; **D,** seeds.

MUSTARD FAMILY — CRUCIFERAE

HARE'S-EAR MUSTARD *Conringia orientalis* (L.) Dumort.

Other name   Rabbit's-ear.

Description   Annual or winter annual, spreading by seeds, the entire plant glabrous and slightly succulent; stems 15 cm to 6 dm high, simple or little branched; leaves bluish-green, 50-125 mm long, large, broad, entire, glabrous; stem leaves alternate, clasping by earlike basal lobes; flowers terminal, about 6 mm across, petals creamy white; pods attached to the elongated flowering stem by spreading stalks 6-12 mm long, pods 75-125 mm long when mature, narrow, 4-angled and square in cross section, narrowed at the tip to a short beak; seeds 2.5 mm long, oblong, brownish, with a small, distinct whitish projection at the lower end, seed surface coarsely granular. Flowering from May to August but mainly in June.

Origin   Europe.

Distribution in Canada   Occurs in all provinces and reaches its greatest abundance in the Prairie Provinces, particularly Saskatchewan.

Habitats   Fields, gardens, waste places, along railroads, and roadsides. Mostly in waste places in Eastern Canada and in grainfields in Western Canada.

Notes   Seeds of this plant may cause poisoning when fed in grain.

Similar plants   Bird rape, *Brassica campestris,* also, has clutching glabrous leaves, but differs from hare's-ear mustard in having yellow flowers, pods rounded in cross section, and rounded seeds. Hare's-ear mustard has whitish flowers, pods square in cross section, and oblong seeds.

Young plants of hare's-ear mustard have a whitish bloom and resemble cabbage seedlings.

Hare's-ear mustard, **Conringia orientalis: A,** plant; **B,** pods; **C,** cross section of a pod; **D,** seed.

TUMBLE MUSTARD *Sisymbrium altissimum* L.

Description   Annual or winter annual, often breaking at the base at maturity and then freely distributed by wind; stems 3-12 dm high, becoming woody as the plant matures, much branched, with spreading simple white hairs mainly at the base, or glabrous; leaves alternate, terminal segment never large and triangular; lower leaves hairy, divided into 6 or 8 pairs of broad segments, these leaves usually die before flowering occurs; upper leaves finely divided into elongate threadlike segments; flowers clustered at the top of the stem, 6 mm wide, petals pale yellow, longer than the sepals; pods wide-spreading from the elongated flowering stem, on stalks about 6 mm long; pods not broader than their stalks, 5-10 cm long, stiff, glabrous, many seeded, beak inconspicuous; seeds about 1 mm long, oblong, irregularly angled, yellow, reddish yellow or greenish olive, appearing greasy, the immature root plainly evident. Flowering throughout the summer.

Origin   Europe.

Distribution in Canada   Occurs in all provinces and is particularly abundant on the open prairies of the west.

Habitats   Grainfields, grasslands, roadsides, along railways, and waste places. Like many other mustards this plant is found mostly along roadsides and in waste places in Eastern Canada and is an aggressive farm weed in the west.

Similar plants   Tumble mustard should not be confused with tall hedge mustard, *Sisymbrium loeselii* L., which has pods 25-40 mm long on stalks about 12 mm long, stalks not as broad as the pods, and a usually triangular terminal leaf segment.

Tumble mustard, **Sisymbrium altissimum: A,** plant; **B,** mature pods;
C, seeds.

MUSTARD FAMILY — CRUCIFERAE

TALL HEDGE MUSTARD *Sisymbrium loeselii* L.

Other name Loesel's hedge mustard.

Description Annual or winter annual; stems to 12 dm high, usually coarse, strongly branched, hairy at least below with prominent downward-pointing hairs, sometimes many stems from a single root; leaves coarsely lobed, the lower and middle leaves with 2-4 pointed lobes on each side, these lobes toothed on the side away from the base of the leaf and generally entire on the other, terminal lobe triangular and larger than the side lobes, basal leaves withered at flowering time; petals about twice as long as the sepals, bright yellow; pods on slender stalks about 12 mm long; pods narrow but broader than their stalks, about 25-40 mm long, many seeded, beak of pod very short; seeds in a single row, oblong, very small, about 0.8 mm long and half as broad, yellow brown, surface wavy under high magnification, shiny. Flowering in early summer.

Origin Europe.

Distribution in Canada Tall hedge mustard occurs in all the provinces from Quebec west, and is most common in Saskatchewan. Serious infestations occur in Saskatchewan south of Bladworth, around Lebret, and south of Balgonie. There are also large infestations in Manitoba: 4 miles north of Ninette, and at Cartwright.

This plant was first collected in Saskatchewan in 1929 and at later dates in the other provinces. Although tall hedge mustard is of comparatively recent introduction, the size of some of the infestations and the vigor of individual plants indicate that it may become a serious pest.

Habitats Field margins, grainfields, fallow, waste places, and roadsides.

Similar plants The triangular terminal lobe of the leaves of tall hedge mustard and the downward-pointing hairs on the lower part of the stem are sufficient for separation from flixweed and other mustards that have similar pods.

Hedge mustard, *Sisymbrium officinale* (L.) Scop., is a plant of European origin occurring in the eastern provinces and British Columbia largely as a garden and waste-place weed. It has coarsely divided leaves and is readily separated from the other *Sisymbrium* species by the following characteristics: its pods may be hairy (or glabrous in var. *leiocarpum* DC.); pods little more than 12 mm long on stalks 1.5 mm long; the mature pods taper from base to tip; pods and stalks erect and appressed to the main stem. Hedge mustard should not be mistaken for black mustard (page 92).

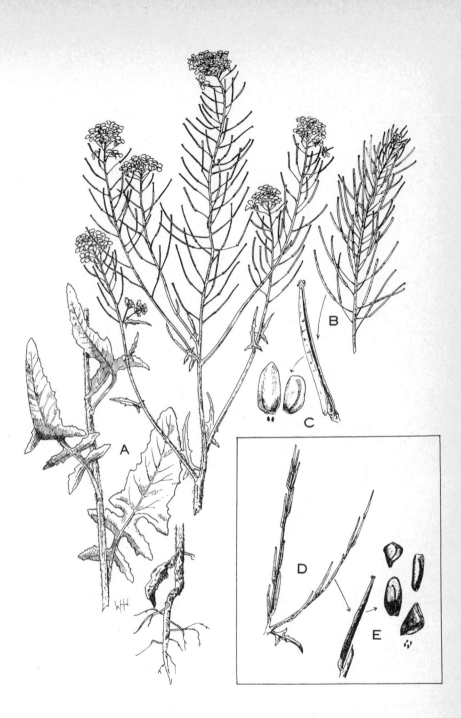

Tall hedge mustard, **Sisymbrium loeselii: A,** plant; **B,** group of pods; **C,** single pod and seeds. Hedge mustard, **Sisymbrium officinale: D,** group of pods; **E,** single pod and seeds.

FLIXWEED *Descurainia sophia* (L.) Webb

Other name   Formerly known by the scientific name *Sisymbrium sophia* L.

Description   Annual or biennial, the whole plant grayish green due to minute branched hairs; stems to 9 dm high, branched above; leaves alternate, all leaves two or three times divided into very narrow segments; flowers very small, clustered at the top of the stem, petals pale yellow and not longer than the sepals; pods spreading from the elongated flowering stem on stalks usually just over 6 mm long; pods very narrow but broader than their stalks, about 8 mm long, slightly curved; seeds 1 mm long, oblong, bright orange, in one row in each half of the pod. Flowering throughout the summer.

Origin   Europe.

Distribution in Canada   Occurs in all provinces. One of the most abundant weeds on the Canadian Prairies.

Habitats   Grainfields, gardens, roadsides, along railways, and waste places.

Similar plants   Flixweed has two close relatives from which it may easily be distinguished. Green or short-fruited tansy mustard, *Descurainia pinnata* (Walt.) Britt. var. *brachycarpa* (Richards.) Fern., is a greenish plant with leaves divided into fine segments. It differs from flixweed in having glandular hairs, seeds in two rows in each half of the pod, and pods about 12 mm long on stalks of nearly the same length.

Gray tansy mustard, *Descurainia richardsonii* (Sweet) O. E. Schulz, is a grayish plant with leaves less divided than those of flixweed. It is rather readily distinguished from flixweed and green tansy mustard as its pods and stalks are erect and close to the main flowering stem. The pods of gray tansy mustard are about 10 mm long and are on stalks 5 mm long.

Both tansy mustards are native to North America and are widely distributed, although practically absent from the Atlantic Provinces. Like flixweed they are most abundant in the Prairie Provinces.

The group of plants, *Sisymbrium* spp., to which tumble mustard belongs is similar to flixweed and to the other plants described above. Species of *Sisymbrium* differ from those in *Descurainia* in having larger flowers, larger leaf segments, and in lacking branched hairs.

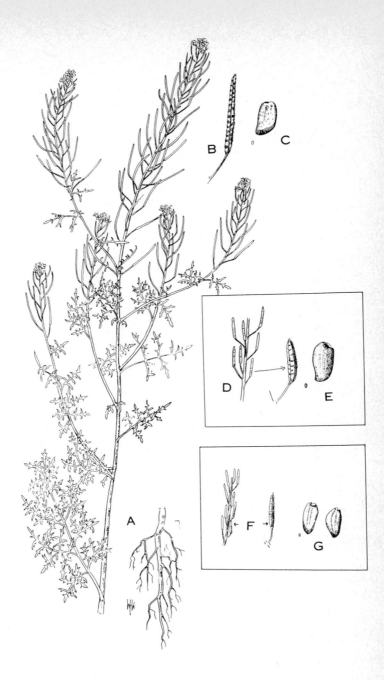

Flixweed, **Descurainia sophia: A**, plant; **B**, pod; **C**, seed. Green tansy mustard, **Descurainia pinnata** var. **brachycarpa: D,** group of pods and single pod; **E,** seed. Gray tansy mustard, **Descurainia richardsonii: F,** group of pods and single pod; **G,** seeds.

MUSTARD FAMILY — CRUCIFERAE

WORMSEED MUSTARD *Erysimum cheiranthoides* L.

Other name   Treacle mustard.

Description   Annual or winter annual, spreading by seeds, taproot short, slender; stems 15 cm to 9 dm high, erect, branching, roughened from 2-branched hairs; leaves alternate, narrow, 25-75 mm long, bright green, margins wavy and with occasional teeth, hairy above and below, hairs predominantly 3-branched; flowers small, about 6 mm across, crowded at first in terminal clusters, separating as the pods mature; petals pale yellow; pods nearly 25 mm long, bluntly 4-angled, usually erect, on slender spreading stalks, beak short; seeds reddish yellow, dull, about 1.3 mm long, variable in shape. Flowering from June to late fall, maturing seed from July.

Origin   Probably introduced from the Old World. Native and weedy throughout Europe except in the southeast; north Asia including Siberia. A naturalized plant in North America, but some botanists think that at least part of the population may be native.

Distribution in Canada   In the agricultural areas of all provinces. Ranges far to the north in the District of Mackenzie and the Yukon.

Habitats   Waste places, cultivated land, grainfields, gardens, railway yards, river shores, and prairie bluffs. In Eastern Canada, more often a weed of waste places than of cultivated land.

Notes   Seedlings or nonflowering plants of wormseed mustard are easily identified by the flat-lying, 3-branched hairs, which can be detected with a low-powered hand lens.

Similar plants   Small-flowered prairie-rocket, *Erysimum inconspicuum* (S. Wats.) MacM., is a stiff, grayish plant with many 2-branched hairs on the leaves, far larger pods than wormseed mustard, and shorter, thicker pod stalks. This native plant is abundant in Western Canada and is now known from most of the eastern provinces.

Tall wormseed mustard, *Erysimum hieraciifolium* L., a European species, was not recognized in Canada until 1954. It is now known from Nova Scotia, Quebec, Ontario, and Saskatchewan. Tall wormseed mustard is particularly abundant in eastern Ontario, where it recently has been found in several new localities. It is a roadside and waste-place weed differing from wormseed mustard in its greater height (to 18 dm), larger flowers, pods on short stalks, pods not spreading as in wormseed mustard but appressed to the stem, and in being biennial and perhaps perennial.

Wormseed mustard, **Erysimum cheiranthoides: A,** plant; **B,** upper part of stem showing pods; **C,** pod; **D,** seeds.

YELLOW ROCKET *Barbarea vulgaris* R. Br.

Other names   Herb barbara, winter cress.

Description   Biennial or perennial, spreading only by seeds, in the first year forming a cluster of glossy dark green leaves; stems to 6 dm high, solitary or several from the same root, branched; leaves glabrous; lower leaves divided into a large terminal lobe and smaller side lobes; stem leaves alternate and decreasing in size towards the top of the stem, stalkless and clasping the stem by basal lobes; the uppermost leaves below the inflorescence very characteristic, rarely divided to the midrib but usually coarsely toothed; flowers about 6 mm across, numerous, petals bright yellow; pods erect, appressed to the stem, strongly overlapping and forming a dense raceme, or in the more common var. *arcuata* (Opiz.) Fries. spreading, not overlapping, and forming an open raceme; pods about 25 mm long on stalks about 3 mm long, containing up to 20 seeds, the slender persistent beak about 2 mm long; seeds metallic grayish brown, about 1.3 mm long, egg-shaped, pitted. Flowering in late May and June.

Origin   Europe.

Distribution in Canada   Known to occur in all provinces except Saskatchewan. Particularly common in Eastern Canada and apparently spreading rapidly. In recent years, it is considered to have spread more than any other weed in Ontario.

Habitats   Meadows, pastures, and roadsides. Especially common on moist rich soils.

Notes   Cultivated in Europe as a potherb.

Similar plants   Yellow rocket is often mistaken for wild mustard. The perennial or biennial habit, earlier flowering, smaller flowers, glabrous and glossy leaves, and clutching stem leaves of yellow rocket should be sufficient for differentiation.

Yellow rocket, **Barbarea vulgaris: A,** plant; **B,** pod; **C,** seeds.

ROUGH CINQUEFOIL *Potentilla norvegica* L.

Other names Upright cinquefoil. The scientific name, *Potentilla monspeliensis* L., appears in some of the older North American texts.

Description Annual, biennial, or short-lived perennial, spreading by seeds; stems 1 or several from the same root, usually 20-50 cm high, robust, hairy, branched; leaves alternate, hairy, green on both sides, each leaf consisting of 3 coarsely toothed leaflets, leaflets oblong or almost rounded; flowers small, about 6 mm across, arranged in open leafy groups; petals 5, separate, pale yellow, not longer than the green sepals; seeds numerous in each flower, about 1 mm long, yellowish or pale brown, longitudinally ridged. Flowering largely in June and July, maturing seed from July.

Origin Native to North America and Europe. Differences between native and introduced forms are not well defined. Probably much of the rough cinquefoil in Eastern Canada is of European origin.

Distribution in Canada Widely distributed across Canada and common in all provinces.

Habitats Grainfields, meadows, pastures, gardens, woods, and waste places. Apparently found in wetter places in the west than in the east.

Similar plants There are a great many species of cinquefoil in Canada. Rough cinquefoil, the most abundant species, is characterized by small flowers, leaves green on both surfaces, and leaves divided into 3 leaflets.

Silvery cinquefoil, *Potentilla argentea* L., is a stout-rooted perennial that has spread aggressively into pastures, open meadows, and lawns on sandy soils in Ontario and western Quebec. It is less common in other provinces, although absent only in Alberta. This cinquefoil has flowers of similar size to those of rough cinquefoil, but its leaves are white woolly below and are divided into 5 or 7 leaflets.

Sulfur cinquefoil, *Potentilla recta* L., a perennial species of European origin, is found in every province and reaches its greatest abundance in Ontario and western Quebec. This species has large deep-yellow or sulfur-yellow flowers about 25 mm across, leaves green on both sides, 5 or 7 leaflets, and prominently ridged seeds with a narrow, winged margin.

The sepals in the cinquefoils appear to be 10 in number, actually because of 5 additional bracts (see *D* in the illustration). This character is useful for distinguishing the group from other similar yellow-flowered plants that usually have 5 sepals.

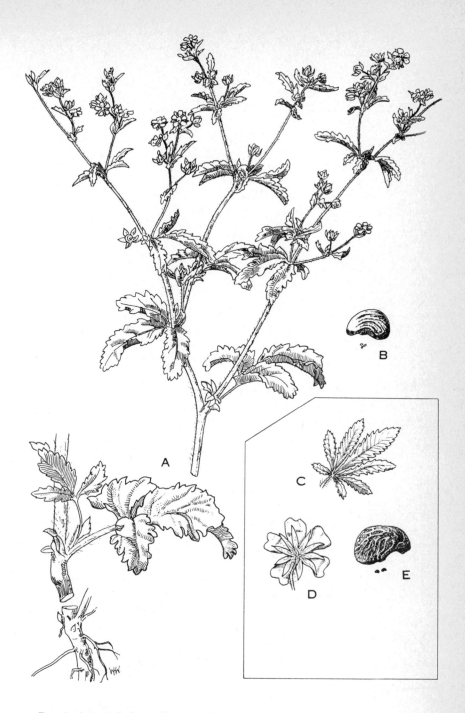

Rough cinquefoil, **Potentilla norvegica: A,** plant; **B,** seed. Sulphur cinquefoil, **Potentilla recta: C,** leaf; **D,** flower showing sepals; **E,** seed.

PULSE FAMILY — LEGUMINOSAE

TUFTED VETCH *Vicia cracca* L.

Other names   Purple-tufted vetch, wild vetch.

Description   Perennial, mat-forming or twining about other plants, spreading by seeds and wiry rootstocks; stems slender and weak, 4-18 dm long; leaves alternate, divided into numerous bristle-tipped leaflets, leaves ending in branched tendrils; flowers bluish purple, nearly 12 mm long, grouped on one side of long stalks arising from the leaf axils; seedpods flat, at maturity splitting into 2 twisted halves, about 18 mm long, light brown, containing several seeds; seeds rounded to oval, ranging from 2.5-3 mm long, reddish brown, sometimes slightly mottled, dull velvety, marked nearly their full length by a whitish- or reddish-brown scar. Flowering from early June to October, ripening seed from early July.

Origin   Europe.

Distribution in Canada   One of the commonest plants of Eastern Canada, but much less common in the four western provinces. In every province.

Habitats   Meadows, pastures, gardens, waste places, grainfields, and cultivated fields. Tufted vetch, related to clover and alfalfa, has fodder value and is probably beneficial in meadows unless better legumes are displaced. Because of its twining habit, it is troublesome on cultivated land and in grainfields.

Similar plants   There are about 10 species of vetch in Canada. Tufted vetch is rather easily recognized by the numerous bluish flowers borne on one side of a long flowering stalk.

Narrow-leaved vetch, *Vicia angustifolia* Reichard, introduced from Europe, is found in much the same habitats as tufted vetch in the eastern provinces and British Columbia. It is an annual, and otherwise differs from tufted vetch in having fewer and larger flowers grouped in the axils of leaves and not on a long stem, and by its black pods, about 40 mm long. Four-seeded vetch, *Vicia tetrasperma* (L.) Moench, also of European origin, is found in Eastern Canada and British Columbia. This annual weed is unlike the other species mentioned here in the following respects: small flowers, about 6 mm long, in groups of 1-6 on slender stems; pods 12 mm long and containing at most 4 very small seeds.

Another legume, black medick, *Medicago lupulina* L., is rather weedy in fields and gardens. It is a yellow-flowered annual, with leaves consisting of 3 hairy leaflets, and with the central leaflet on a longer stalk than the lateral pair.

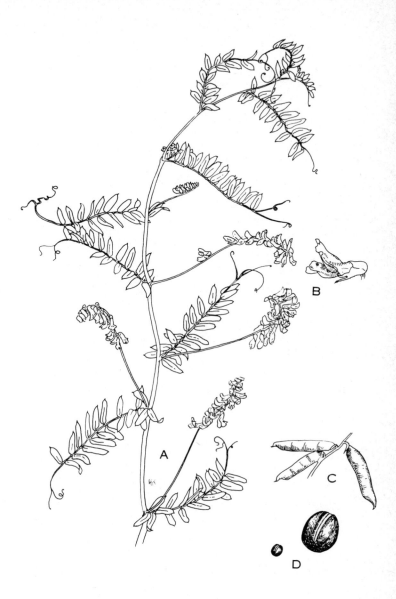

Tufted vetch, **Vicia cracca: A,** plant; **B,** flower; **C,** part of stem with mature pods; **D,** seed.

LEAFY SPURGE *Euphorbia esula* L.

Other names The leafy spurge that is weedy in North America has also been referred to *Euphorbia virgata* Waldst. & Kit., and there still remains the possibility that the very variable plant considered to be *Euphorbia esula* includes the former species and perhaps others. Unfortunately the vast and widely distributed assemblage of species related to *Euphorbia esula* is not well understood, even in Eurasia, and it will be some time before the scientific name of our leafy spurge is finally settled.

Description Perennial, spreading by seeds but largely by persistent vertical and horizontal underground rootstocks on which shoot buds are produced, all parts of the plant containing a milky juice; stems in clumps and often forming dense and extensive stands, erect, glabrous, up to 7 dm high; leaves of the stem spirally arranged, entire, glabrous, 4-12 mm broad, 25-75 mm long, narrowed to the base; flowers inconspicuous, greenish yellow, either grouped in flattish clusters at the tips of the stem with a whorl of leaflike bracts below or scattered along the stem; the flowers are inserted among pairs of floral bracts, bracts roundish but pointed at the tip; flowers very complex, a number of male flowers with a single stamen and a single female flower with a capsule being grouped together; capsules 3-celled, each cell with 1 seed; seeds about 2.5 mm long, surface finely netted, yellowish, grayish, brown or brownish mottled, with a dark line down one side, usually with a yellow appendage at the base. Flowering mainly in June.

Origin Europe.

Distribution in Canada Leafy spurge, now well established and spreading rapidly in Western Canada, is a major threat to prairie lands. Southern Ontario and southeastern Quebec have a number of localized infestations. Newfoundland appears to be free of this weed.

Habitats Grainfields, meadows, pastures, roadsides, and waste places. Penetrating into native rangeland in Western Canada. Found on a wide variety of soils from sands to clays.

Notes The milky juice of leafy spurge and other spurges may cause severe skin rashes in humans. Spurges are poisonous to most livestock, although sheep apparently eat leafy spurge without harmful effect.

Similar plants Cypress spurge (see page 114) is closely related to leafy spurge but differs in having much shorter and narrower, almost bristlelike stem leaves and, after flowering, in its bushy habit resulting from the development of very leafy branches.

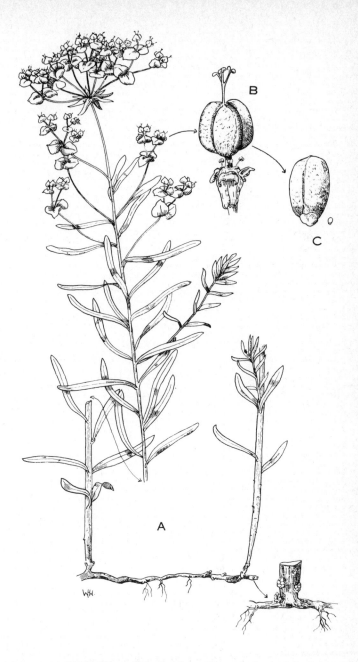

Leafy spurge, **Euphorbia esula: A,** plant; **B,** a single female flower with several male flowers at its base; **C,** seed.

SPURGE FAMILY — EUPHORBIACEAE

CYPRESS SPURGE *Euphorbia cyparissias* L.

Other names   Graveyard spurge or graveyard weed.

Description   Perennial, with a milky juice in all parts, spreading by seeds and thick rootstocks, forming large patches; stems about 3 dm high; stem leaves very narrow, rarely over 1.5 mm wide, 25 mm or less in length, dropping off early, leaves of the branches bristlelike and even narrower than the stem leaves; flowers similar to those of leafy spurge, floral bracts without the sharp tip found in leafy spurge; seeds rarely produced, similar to those of leafy spurge, about 2.5 mm long, yellow, bluish gray, or purplish red. Flowering mainly in June.

Origin   Europe. Introduced into Canada as an ornamental.

Distribution in Canada   Common in southern Ontario and eastern Quebec. Known from all provinces except Alberta.

Habitats   Along roadsides often as an escape from cemeteries, waste places, pastures, and wood openings.

Notes   Cypress spurge is a poisonous plant. Cases of dermatitis in people who have handled the plant have been reported from Eastern Canada.

Fortunately seed production is very rare in cypress spurge; only nine of the many infestations in Canada are known to produce seed. The infestations setting seed are extensive and contain many patches, while the sterile stands are often confined to a single, slowly spreading patch.

Similar plants   See the description of leafy spurge. After flowering, cypress spurge is a very distinctive plant, the stem leaves drop and all the leaves are then confined to the upward pointing side branches, which are so densely leafy that the general appearance of the plant at this stage suggests an evergreen seedling.

Some of the spurges weedy in Canada are annual plants. One of the most common of these annuals is sun spurge or wartweed, *Euphorbia helioscopia* L., shown on the opposite page. Sun spurge is easily recognized by its spoon-shaped and minutely toothed leaves, the whorls of 5 broad leaflike bracts below the inflorescence, and the coarsely netted seeds. Sun spurge occurs in all provinces but is common only in Eastern Canada.

There are also prostrate annual spurges. These plants have small opposite leaves, flowers similar to the other spurges, a usually prostrate habit, and the milky juice common to all spurges. At least one species is found in every province in cultivated land, gardens, and along railways. The two commonest species, particularly in the west, are *Euphorbia serpyllifolia* Pers. and *Euphorbia glyptosperma* Engelm.

Cypress spurge, **Euphorbia cyparissias: A,** plant; **B,** seed. Sun spurge,
Euphorbia helioscopia: C, plant; **D,** seed.

CASHEW FAMILY — ANACARDIACEAE

POISON-IVY *Rhus radicans* L.

Other names   Poison-ivy is sometimes called poison oak, a name properly applied to a plant of the Pacific Coast, *Rhus diversiloba* Torr. & Gray. *Rhus toxicodendron* L., a scientific name once used for poison-ivy, is now reserved for a species of the Eastern United States.

Description   A woody perennial spreading by seeds and sucker shoots; stems sometimes climbing to 60-90 dm on trees, etc., and developing aerial roots, or trailing along the ground, or upright, and 15 cm to 9 dm high; leaves alternate, consisting of 3 stalked leaflets, middle leaflet with the longest stalk, leaflet margins very variable, leaves purplish red in spring, bright green and glossy in summer, brightly colored in autumn; flowers small, whitish green, often absent, clustered in the leaf axils; fruits not conspicuous until leaf fall, roundish, nearly 6 mm in diameter, 1-sided, dull white in late summer, yellowish brown by the next spring; winter buds brown woolly and without scales. Flowering June to August, fruiting from mid-July.

Origin   Native to North America.

Distribution in Canada   Occurs in all provinces with the possible exception of Newfoundland. More abundant from Quebec City to the Great Lakes than elsewhere in Canada.

Habitats   Woodlands, beaches, fencerows, roadsides, and rocky and waste places.

Notes   The characteristic skin blistering caused by a poisonous principle in poison-ivy results from direct contact with the plant in any season or from handling shoes, garden tools, or other articles, contaminated perhaps years before. Household pets can also carry the poisonous substance. Compost containing poison-ivy roots is often a source of trouble.

Similar plants   Many plants are mistaken for poison-ivy. Virginia creeper differs in having 5 leaflets and bluish fruits. Manitoba maple sometimes with 3 leaflets, has opposite leaves and winged fruits. Hog peanut has 3 leaflets, but is not woody, and has larger flowers than poison-ivy.

Poison sumac, *Rhus vernix* L., found occasionally in swamps in southern Ontario and western Quebec, contains the same rash-producing substance as poison-ivy. Poison sumac is a tall shrub to 60 dm high, with alternate leaves consisting of 3-6 pairs of entire-margined leaflets and a single terminal leaflet. White ash has similar leaves, which, however, are opposite. Staghorn sumac, *Rhus typhina* L., has bright red fruits, sharply toothed leaflets, and is not poisonous.

A publication on poison-ivy is available from Agriculture Canada, Ottawa, Ontario.

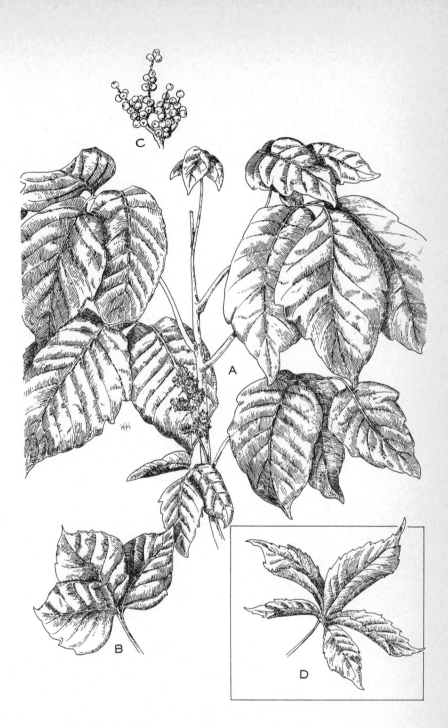

Poison-ivy, **Rhus radicans: A,** plant; **B,** leaf; **C,** cluster of fruits. Virginia
creeper, **Parthenocissus quinquefolia,** Vine family—Vitaceae: **D,** leaf.

BUCKTHORN FAMILY — RHAMNACEAE

EUROPEAN BUCKTHORN *Rhamnus cathartica* L.

Other name   Common buckthorn.

Description   Many stemmed bush 18-45 dm high or rarely a small tree to 90 dm high, spreading only by seeds, regenerating quickly after cutting and burning, branches almost at right angles to the stem and opposite or nearly opposite, many of the branches ending in a thorn; leaves 25-50 mm long, usually almost opposite or sometimes alternate, as *A* in illustration, the margins with small teeth, leaf veins few, veins curved towards the leaf tip and more or less parallel; flower parts in fours, flowers green, rather small and inconspicuous, on slender stalks, solitary or grouped in leaf axils; fruits green at first then black on ripening, roundish, at maturity about 6 mm across, 3- or 4-seeded. Flowering in early June, fruits forming by late June and maturing by September.

Origin   Europe. Introduced to North America as a hedge or windbreak plant. Seeds spread by birds.

Distribution in Canada   Very common in eastern Ontario, particularly in Ontario, York, Durham, and Victoria counties, and common in eastern Quebec and the Annapolis Valley in Nova Scotia. Apparently the only provinces free of European buckthorn are British Columbia, Alberta, and Newfoundland.

Notes   European buckthorn is not an important weed, but it should be known to farmers because it is the alternate host for crown rust of oats. The black spores of crown rust that have overwintered on oat stubble cannot infect oats, but they do infect buckthorn, on which the cluster cup spores of crown rust are produced. These spores attack new crops of oats and may be responsible for very serious losses.

Similar plants   Alder or glossy buckthorn, *Rhamnus frangula* L., also introduced from Europe, is similar to European buckthorn, but differs in having spineless branches, winter buds without scales, and densely hairy (buds scaly and glabrous in European buckthorn), entire leaves, more leaf veins, and flower parts in fives. Alder buckthorn is only found east of the Great Lakes in Canada. It is not an alternate host for crown rust of oats.

Buckthorns are easily detected in late fall, as leaves are retained longer than on other woody shrubs.

European buckthorn, **Rhamnus cathartica: A,** branch; **B,** leaf; **C,** tip of branch showing a thorn; **D,** flower; **E,** clusters of fruits after leaves have fallen.

MALLOW FAMILY — MALVACEAE

COMMON MALLOW *Malva neglecta* Wallr.

Other name   Cheeses.

Description   Annual to short-lived perennial, spreading by seed; stems prostrate to semierect, 10-60 cm long; leaves alternate, stalk much longer than the roundish, shallow, toothed blade; flowers from one to several on stalks in leaf axils, flowers with 5 separate, whitish to pale lilac petals about 12 mm long; each flower giving rise to a group of 12-15 distinct nutlets (carpels) that form a disk about 6 mm in diameter; nutlets rounded and smooth on the back, thin-walled and covered with minute hairs, each nutlet containing one dark brown seed; seed finely roughened, circular in outline, and flattened, about 1.5 mm in diameter. Flowering from May to October.

Origin   Europe.

Distribution in Canada   Found in all provinces except Saskatchewan. Most common in the settled areas of Quebec, Ontario, and British Columbia.

Habitats   Gardens, farmyards, roadsides, waste places, and occasionally in cultivated fields.

Similar plants   Common mallow is easily mistaken for round-leaved mallow, *Malva pusilla* Sm., also introduced from Europe. Round-leaved mallow can be differentiated from common mallow by its smaller petals, about 6 mm long, and the network of roughened ridges on the back of each nutlet. It is found in all provinces except Newfoundland, but is common only in the Prairie Provinces where it virtually replaces common mallow.

Several other species of mallows that grow in gardens have escaped to become weeds. The most common of these is musk mallow, *Malva moschata* L. It can be distinguished from common mallow and round-leaved mallow by larger petals, which are at least 25 mm long, and by 5-parted upper stem leaves with divisions cleft into narrow segments. Musk mallow is locally common in Newfoundland, Prince Edward Island, Nova Scotia, New Brunswick, Quebec, Ontario, and British Columbia.

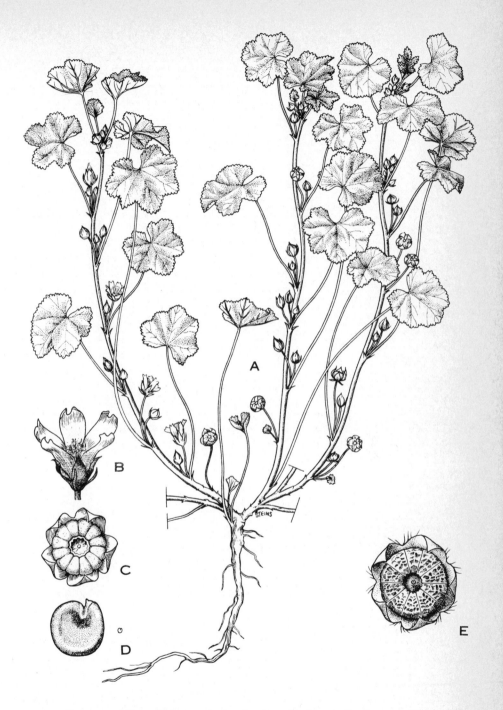

Common mallow, **Malva neglecta: A,** plant; **B,** flower; **C,** disk of nutlets; **D,** seed. Round-leaved mallow, **Malva pusilla: E,** disk of nutlets.

ST. JOHN'S-WORT FAMILY — HYPERICACEAE

ST. JOHN'S-WORT *Hypericum perforatum* L.

Other names   Goatweed, klamath weed.

Description   Perennial, spreading by seeds and by shoots from underground runners; stems 3-9 dm high, erect, much branched, rust-colored, and somewhat woody at base, green above, 2-ridged, without hairs; leaves opposite, sessile, entire, not over 25 mm long, narrow, dotted with transparent glands, sometimes also with black dots; flowers numerous and showy, about 18 mm across, the 5 separate yellow petals with black dots around the edges, petals twice as long as the sepals, stamens very evident and grouped into 3 clusters; seedpod rusty brown, about 6 mm long, with 3 persistent styles nearly as long as the pod, pods at maturity separating into 3 sections, each with many seeds; seeds 0.75 mm long and half as wide, dark brown, pitted, round in cross section, short pointed at one end. Flowering June to September.

Origin   Europe.

Distribution in Canada   Occurs in Eastern Canada and British Columbia. Apart from a report of its presence at Winnipeg, it is apparently absent between Manitoulin Island, in Ontario, and British Columbia. More than 0.8 million hectares of rangeland in the Western United States have been invaded by this weed, and it is gaining a foothold in the grazing lands of British Columbia.

Habitats   Rangelands, pastures, meadows, roadsides, and waste places; often on sandy or gravelly soils.

Notes   Owing to the abundance of this weed in Australia, scientists investigated the possibility of controlling it with insects from Europe. Effective control was obtained with three insect species that by appropriate tests were shown to feed exclusively on plants of the St. John's-wort family. These insects have been released in the Western United States, and, in 1951, in Canada.

St. John's-wort contains a toxic substance that affects white-haired animals when they are exposed to strong sunlight after having eaten the plant. Poisoned animals rarely die but they suffer severe irritation and loss of weight.

Similar plants   When plants of common St. John's-wort are held up to the light, small transparent dots can be seen on the leaves. Native species related to this weed share this character but are usually found in wettish habitats and are rarely weedy.

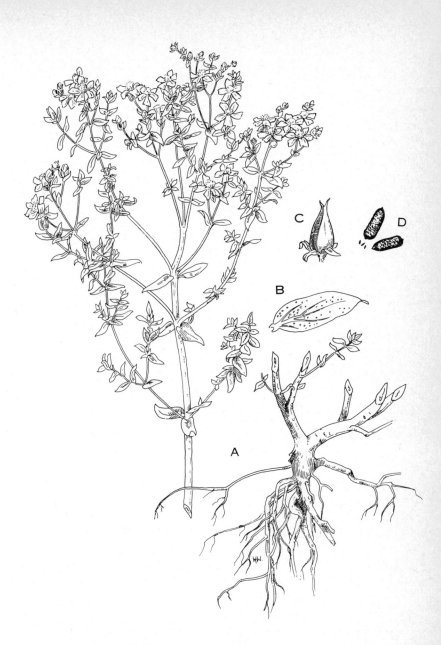

St. John's-wort, **Hypericum perforatum: A,** plant; **B,** leaf showing transparent dots; **C,** pod; **D,** seeds.

EVENING-PRIMROSE FAMILY — ONAGRACEAE

YELLOW EVENING-PRIMROSE *Oenothera biennis* L.

Other name   Common evening-primrose. Yellow evening-primrose is a variable species and consists of several races that are sometimes considered to be varieties or even separate species.

Description   Biennial with a large fleshy taproot, in the first year producing a flat-lying rosette, spreading by seeds; stem erect, 6-18 dm high, often unbranched, usually hairy, becoming very coarse and woody; rosette leaves stalked, leaves of the stem sessile, alternate, leaf margins sparsely toothed; flowers in leaf axils, large, showy, with 4 yellow petals that are more than 12 mm long; seedpods 25-40 mm long, hairy, not winged, opening from the top into 4 cells each with many seeds; seeds 1.5 mm long, dull reddish brown, very irregular in shape, 4- or 5-sided and with acute winged angles. Flowering from July to September.

Origin   North America. Introduced to Europe in the seventeenth century and now widely established in Europe and Asia.

Distribution in Canada   Occurs in all provinces; more common in the east than the west.

Habitats   Meadows, pastures, waste places, and roadsides.

Similar plants   Evening-primroses are easy to recognize with their 4 petals and distinctive seeds and seedpods. Of the several species native to Canada, yellow evening-primrose occurs most abundantly as a weed. Two other species deserve mention. Sundrops, *Oenothera perennis* L., is a perennial found in waste places and on pastures in Eastern Canada. It is shorter than yellow evening-primrose, has yellow petals less than 12 mm long, and strongly winged pods that are club-shaped at the upper end. White evening-primrose, *Oenothera nuttallii* Sweet, is a perennial, common and persistent in the sandy lands of the western provinces. Large white flowers and stems with shining, white, shredding bark are characteristic of this plant.

Fireweed or great willow-herb, *Epilobium angustifolium* L., is closely related to the evening-primroses from which it differs in having pinkish flowers and long narrow pods containing many tiny seeds, each provided with a tuft of whitish hairs. Fireweed is a tall perennial plant with creeping rootstocks, narrow willowlike leaves, and long pyramidal spikes of flowers. It is abundant throughout Canada and often persists after forest fires or in recent clearings.

Yellow evening-primrose, **Oenothera biennis: A,** plant; **B,** seeds.

PARSLEY FAMILY — UMBELLIFERAE

POISON HEMLOCK *Conium maculatum* L.

Other names   Deadly hemlock, hemlock, poison parsley, poison stinkweed, snakeweed.

Description   Biennial, often with a disagreeable odor particularly when crushed, completely glabrous, spreading by seeds; taproot whitish, sometimes branched, thick, 20-25 cm long; stems to 18 dm tall, erect, branched, smooth, usually covered with purplish spots; leaves 10-30 cm long, alternate, soft, dark glossy green, divided into finely cut leaflets, which give the plant a lacy appearance, veins of the leaflets ending in a short colorless bristle tip; flowers small, white, in large loose clusters; clusters with a circle of narrow bracts at the base; fruits composed of 2 seeds (mericarps), each seed with 5 prominent wavy ridges running from top to bottom. Flowering from June to August.

Origin   Europe, Asia, and North Africa. Probably introduced from Europe.

Distribution in Canada   A rather rare plant in Canada. Found in Eastern Canada and British Columbia. Practically absent from the Prairie Provinces.

Habitats   Dry ground, usually on field borders, roadsides, waste places, and dry ditches.

Notes   This plant is very poisonous but its rarity in Canada and its habitats make it very unlikely to be the cause of serious loss of livestock. Dangerous qualities are due to the presence of several narcotic alkaloids.

Historically, poisonous draughts prepared from this plant were used by the ancient Greeks to execute political enemies.

Similar plants   Poison hemlock is often mistaken for spotted water-hemlock (page 128). The clusters of fleshy roots, cross partitions at the stem base, broader leaflets, and wetter habitats of spotted water-hemlock distinguish these plants.

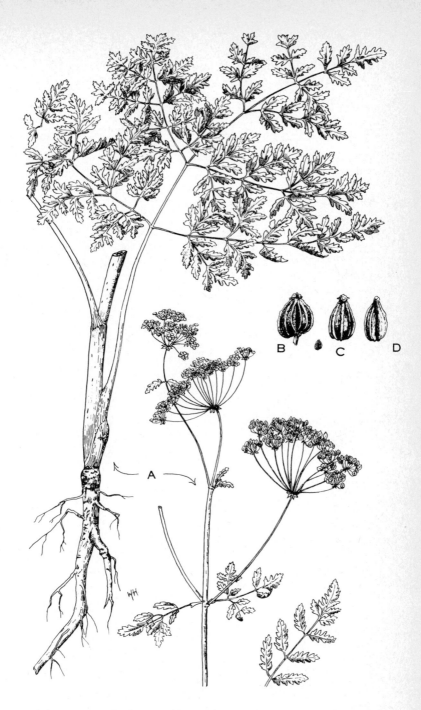

Poison hemlock, **Conium maculatum: A,** plant; **B,** a mature fruit consisting of 2 seeds; **C,** outer surface of seed; **D,** inner surface of seed.

PARSLEY FAMILY — UMBELLIFERAE

SPOTTED WATER-HEMLOCK *Cicuta maculata* L.

Other names   Beaver poison, musquash-root, poison parsnip, spotted cow-bane, spotted hemlock.

Description   Perennial, spreading by seeds; roots thick and fleshy, in clusters around the stem base, containing a yellow secretion; stems erect, 9-18 dm high, hollow, streaked with purple, the somewhat swollen base with cavities separated by cross partitions of solid tissue easily seen when the stem is split lengthwise at the base just above the roots; leaves 10-30 cm long, alternate, glabrous, each leaf divided into 3 parts, each part again divided into leaflets; leaflets coarsely and sharply toothed; flowers white, small, in spreading clusters (umbels), usually without bracts below each main cluster; fruits light to dark brown, oval or ob-long, separating when mature into 2 seeds, each seed with 5 rounded ribs on one side, the other side flattened. Flowering from June to August.

Origin   Native to Canada.

Distribution in Canada   Water-hemlocks as a group occur across Canada and north to the District of Mackenzie and the Yukon.

Habitats   Wet places: swamps, ditches, stream and river banks, meadows, and pastures.

Notes   Of all Canadian plants, the water-hemlocks are the most poisonous to both human beings and livestock. The poisonous substance, cicutoxin, is found in the whole plant, but particularly in the roots. Animals are poisoned usually in the spring, when the roots are readily pulled out of the soft ground.

Similar plants   Several species of water-hemlock occur in Canada. These closely related plants differ in minor ways from the plant illustrated. Bulbous water-hemlock, *Cicuta bulbifera* L., is readily recognized by the small bulblets at the angles formed by leaf and stem.

Other white-flowered plants belonging to the parsley family occur in the same habitats as the water-hemlocks and are often mistaken for them. The doubly compound leaves and thick roots of the water-hemlocks differentiate it from water-parsnip, *Sium suave* (page 130), which has once-pinnate leaves and slender roots. Other plants similar to water-hemlock lack cross partitions at the base of the stem. Freshly cut surfaces of the underground stem and roots of the water-hemlocks will usually show a yellow oily secretion not present in other similar plants.

128

Spotted water-hemlock, **Cicuta maculata: A,** plant; **B,** outer and inner
surfaces of a seed.

PARSLEY FAMILY — UMBELLIFERAE

WATER-PARSNIP *Sium suave* Walt.

Description   Perennial, spreading by seeds, roots slender, stems 6-18 dm high, sometimes with poorly defined cross partitions at the base; leaves alternate, 10-40 cm long, those of the stem consisting of 7-17 saw-edged leaflets, arranged in pairs along the leaf stem, and with a leaflet at the tip, leaflets very variable in width; leaves growing below water are much more finely divided; flowers white, in clusters, with several narrow bracts below the clusters; fruit oval, splitting into 2 seeds, each seed with 5 ribs. Flowering from July to October, maturing seed in September.

Origin and distribution in Canada   Native to North America. From Newfoundland to British Columbia and in the north to James Bay and the District of Mackenzie.

Habitats   Swamps, low marshy ground, river and lake shores, and in shallow water.

Notes   Water-parsnip has been reported as poisonous to stock. Other writers state that the plant is harmless. Feeding trials by one experimenter did not produce symptoms in cattle. Thorough investigation is needed.

Similar plants   The pinnate leaves of water-parsnip differentiate it from spotted water-hemlock and poison hemlock.

WILD PARSNIP *Pastinaca sativa* L.

Description   Biennial, spreading by seeds; taproot thick, mostly unbranched; stems coarse, erect, 9-15 dm tall; leaves alternate, arranged in pairs along a simple axis and with an odd leaflet at the tip, leaflets with an irregular outline and somewhat coarsely lobed; flowers small, yellow, in clusters, bracts below each main cluster usually lacking but sometimes 1 or 2; fruits oval, separating at maturity into 2 strongly flattened seeds with flat sides together, each seed with conspicuous dark lines (oil tubes) on the broad surface. Flowering from May to October.

Origin and distribution in Canada   This European plant now occurs in every Canadian province and is particularly common in eastern Ontario and adjacent Quebec.

Habitats   Roadsides, railway embankments, old fields, meadows, river banks, and ditches.

Notes   Contact with the leaves may produce a severe skin irritation in some individuals.

Similar plants   The yellow flowers, flattened and winged seeds, and leaflet shape distinguish wild parsnip from water-parsnip.

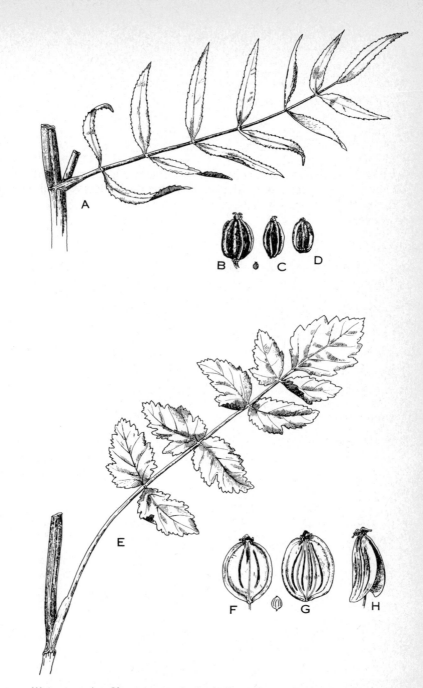

Water-parsnip, **Sium suave: A,** leaf; **B,** a mature fruit consisting of 2 seeds; **C,** outer surface of seed; **D,** inner surface of seed. Wild parsnip, **Pastinaca sativa: E,** leaf; **F,** inner surface of seed; **G,** outer surface of seed; **H,** fruit splitting into 2 seeds.

PARSLEY FAMILY — UMBELLIFERAE

WILD CARROT *Daucus carota* L.

Other names   Bird's nest, Queen-Anne's-lace.

Description   Annual or biennial, spreading by seed; taproot slender, whitish, and woody; stems rough bristly, ridged, branched near the top, 3-9 dm tall; leaves alternate, finely divided, hairy, attached to the stem by a sheathing base; flowers white, numerous, in a flat circular flower cluster or umbel, which becomes concave as the fruits mature producing "bird's nests," usually with a single purple flower in the center of the umbel, umbels surrounded by a whorl of green, finely cut bracts; fruits about 3 mm long, separating into 2 seeds, each seed with several rows of spiny bristles on one side, light brown. Flowering from mid-July to late-August, maturing seeds from late August to October.

Origin   Eurasia.

Distribution in Canada   Abundant in the old settled regions of Ontario and Quebec and still spreading in this area. Not common in the Maritime Provinces, and apparently absent from the Gaspe in Quebec and from Newfoundland. Present in southern coastal British Columbia. There is no definite evidence that the plant occurs in the Prairie Provinces.

Habitats   Pastures, old meadows, roadsides, and waste places. Found on a wide range of soil types.

Notes   Although the wild carrot is supposedly an ancestor of the cultivated carrot, botanists differ as to its actual role in the development of the garden vegetable. However, the wild and cultivated carrots are now distinct strains and it is wrong to suppose that if the wild carrot were transferred to the garden it would yield an edible root or that the cultivated carrot might occur spontaneously as a weed. The wild carrot may cross with the cultivated carrot, which leads to difficulties in commercial seed production.

Similar plants   Caraway, *Carum carvi* L., is sometimes mistaken for wild carrot, but it is readily differentiated from it by the following characters: root thicker, brownish, enlarged near the crown, circular ridges below crown; stems and leaves glabrous; flowers in open umbels; only 1 to 3 narrow bracts below the umbel; seeds ribbed but without bristles, scented, with caraway taste; earlier flowering, in flower by mid-June, and maturing seed by late July.

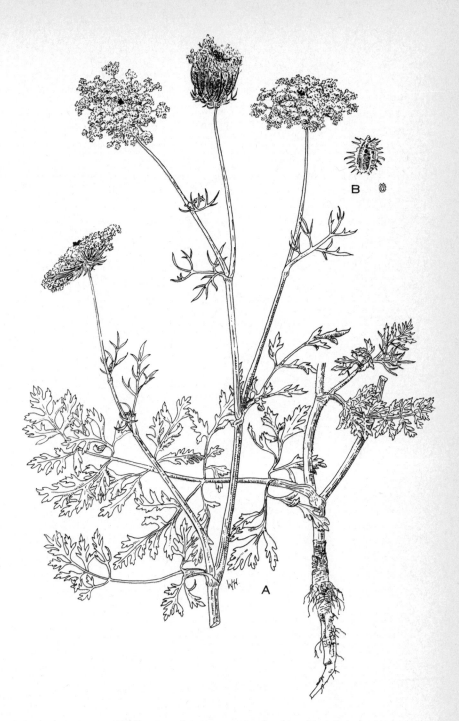

Wild carrot, **Daucus carota: A,** plant; **B,** seed.

MILKWEED FAMILY — ASCLEPIADACEAE

COMMON MILKWEED *Asclepias syriaca* L.

Description   Perennial with a white, sticky juice, spreading by seeds and by thick, fleshy creeping roots producing buds that develop into shoots; stems usually 6-12 dm high, usually several together, unbranched, stout; leaves opposite, oblong, entire, on a short thick stalk, rounded or, less commonly, tapering to the base; leaves 5-10 or rarely 13 cm broad, undersurface velvety from fine hairs, upper surface except of young leaves almost without hairs; flowers in ball-like clusters at the top of the stem and in the axils of leaves, each flower with a stalk, showy, purplish or rarely white, fragrant, only a few of the numerous flowers developing into fruits; fruits (follicles) gray, hairy, and covered with soft projections, fruits 8-10 cm long, about 25 cm broad, containing many seeds; seeds flat, with a broad margin, brown, about 8 mm long, each seed with a tuft of silky hairs. Flowering in late June and July, producing seed in August and September.

Origin   Eastern North America.

Distribution in Canada   Found in southern Manitoba and in all the eastern provinces except Newfoundland, and reaching highest concentrations in the southern parts of Ontario and Quebec. Abundant on Manitoulin Island, Ontario.

Habitats   Pastures, grainfields, cultivated fields, roadsides, fencerows, and waste places.

Similar plants   Several other milkweeds are native to Canada. None are so weedy as common milkweed and only two of the more abundant species will be mentioned here. Swamp milkweed, *Asclepias incarnata* L., ranging from Manitoba to Nova Scotia, is found in swamps, in ditches, and on shores. It differs from common milkweed in having narrower leaves that are not woolly on the lower surface and its fruits are less than 12 mm broad and without projections. Showy milkweed, *Asclepias speciosa* Torr., occurs in the southern parts of all the western provinces. Although very similar to common milkweed, it differs in having densely woolly flower stalks and fewer and longer flowers.

   The dogbanes, *Apocynum* spp., also have a milky juice and seeds with a tuft of hairs, but are unlike the milkweeds in their small bell-shaped flowers, very narrow fruits, and small slender seeds.

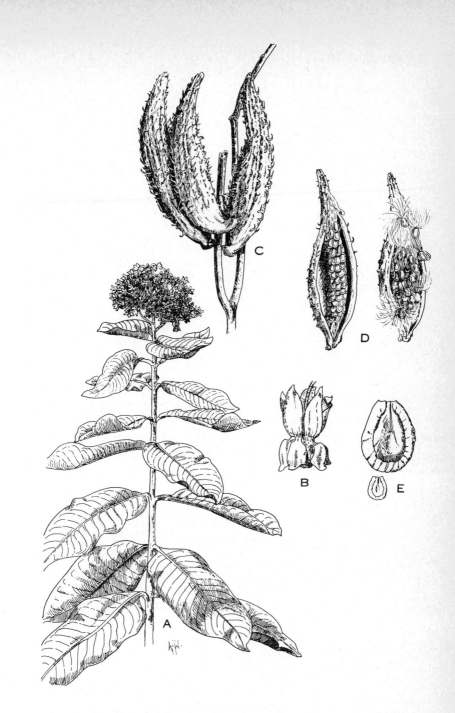

Common milkweed, **Asclepias syriaca: A,** upper part of plant; **B,** flower; **C,** group of fruits; **D,** opening fruit showing seeds; **E,** seed without tuft of silky hairs.

MORNING-GLORY FAMILY — CONVOLVULACEAE

FIELD BINDWEED *Convolvulus arvensis* L.

Other names   European bindweed, small-flowered morning-glory.

Description   Perennial with a deep and extensive root system, spreading by seeds and by shoot buds produced on the roots; roots cordlike, fleshy; stems either prostrate when without a support and then the plant forming dense mats, or stems twining in a counterclockwise direction about other plants, stems sometimes more than 30 dm long; leaves very variable in size and outline, stalked, nearly arrow-shaped, the base with 2 sharp or rounded lobes; leaf margins entire; flowers 25 mm or less across when fully open, solitary or in groups of 2-4, the long flower stems with 2 tiny bracts placed well below the flower; sepals short, not enclosed in 2 larger bracts; seedpods about 3 mm across, containing 1-4 seeds; seeds rather more than 3 mm long, pear-shaped in outline, one side rounded, the other with a rounded central ridge, dull gray-brown, surfaces roughened with grayish warts, basal scar prominent and reddish, seed production rare in many localities. Flowering from the middle of June to September.

Origin   Europe. Now spread over most of the world except the tropics.

Distribution in Canada   In every province with the possible exception of Newfoundland and Prince Edward Island. This persistent weed is particularly troublesome in the southern prairies, and the longer settled areas of Ontario and Quebec.

Habitats   Cultivated land, grainfields, pastures, meadows, roadsides, and waste places.

Similar plants   Hedge bindweed or wild morning-glory, *Convolvulus sepium* L., often mistaken for field bindweed, is very variable in flower color and leaf shape and is divided by botanists into a number of varieties. Some of these varieties are native to Canada and others are considered to be of European origin. This plant is found in every province and sometimes becomes as serious a pest as field bindweed. Hedge bindweed differs from field bindweed in the following ways: much larger flowers, 40-50 mm long, compared with less than 25 mm in field bindweed; flowers and capsules enclosed in 2 large bracts about 18 mm long; flower stalks lacking the 2 tiny bracts placed well below the flowers in field bindweed; and much larger seeds, about 5 mm long, roundish, blackish, with black warts.

Also see "Similar plants" under wild buckwheat (page 42).

Field bindweed, **Convolvulus arvensis: A,** plant; **B,** seed.

DODDERS *Cuscuta* spp.

Description   Annual parasitic flowering plants; stems orange or reddish, threadlike, twining about herbs and shrubs and adhering to these plants by means of suckers; leaves not evident, consisting of minute scales; flowers small, white or cream-colored, usually clustered; seedpods rounded, containing 1-4 seeds; seeds small, 0.8-1.7 mm long, gray to brown, roughened, often nearly round but usually somewhat angled.

Notes   The dodders are distinctive in life cycle and appearance. Their small seeds germinate in the soil and produce slender stems without seed leaves. Unless the slowly rotating stem encounters a host plant within a limited time the dodder seedling withers and dies. If the stem comes in contact with the living stem of a susceptible plant, the dodder twines round it and at numerous points develops suckers, which penetrate the tissue of the host. Food is received through these suckers and the dodder loses all contact with the soil. The dodder spreads rapidly by means of its threadlike stems and, in severe cases, forms dense mats. After a period of growth, clusters of small flowers appear and large amounts of seed are produced. This seed falls to the ground or is harvested with the crop.

Species of dodder   The characters used for identifying the various dodders are so technical that no useful purpose would be served here in describing the individual species. Some 10 species of dodder have been found in Canada, and, of these, the following appear to be those of greatest importance.

Swamp or common dodder, *Cuscuta gronovii* Willd., is a native species ranging from the Maritimes to Saskatchewan. Although usually found on various native plants, swamp dodder sometimes infests crops in low wet fields, but there is no evidence that it will persist for long.

Field dodder, *Cuscuta campestris* Yuncker, a native species, has infested considerable acreages of croplands in Ontario and Quebec. Field dodder persists in agricultural land and is a serious threat.

Large-fruited dodder, *Cuscuta umbrosa* Hooker, formerly called *Cuscuta curta* (Engelm.) Rydb., a native species, is found on important crops in Manitoba and Saskatchewan.

Clover dodder, *Cuscuta epithymum* Murr., introduced from Europe, has been found in Ontario and British Columbia but does not seem to be very persistent.

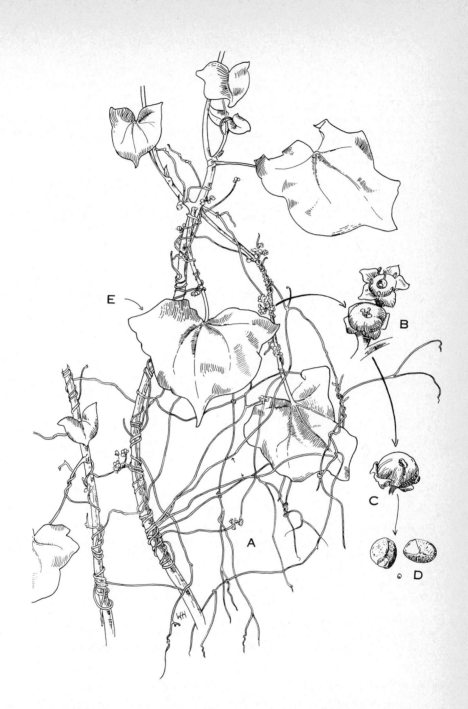

Dodder, **Cuscuta** sp.: **A,** dodder growing on Tartary buckwheat; **B,** flowers; **C,** mature pod; **D,** seeds; **E,** leaf of host plant, Tartary buckwheat.

BORAGE FAMILY — BORAGINACEAE

BLUEWEED *Echium vulgare* L.

Other names   Blue devil, viper's bugloss.

Description   Biennial, forming a flat-lying rosette in the first year, spreading by seeds; taproot long, stout, and black; stems 3-9 dm high, stiff and erect, covered with short hairs and with scattered long stiff hairs, long hairs often with swollen reddish or blackish bases that form conspicuous flecks on the stem; leaves covered with stiff hairs that are sometimes swollen at the base, leaves alternate, entire, 5-15 cm long, the longer leaves in a rosette or at the base of the stem; flowers numerous, arranged on the upper sides of short stems that elongate after flowering; petals large, reddish purple in bud becoming bright blue, very rarely white or pink; seeds in groups of 4, each seed 3 mm long, grayish brown, wrinkled, rounded on one side, angular, and with a central ridge on the other side, pointed at the tip, flattened at the base, with a collarlike margin within which are two small projections and a cavity. Flowering June to August.

Origin   Europe.

Distribution in Canada   Abundant in southern Ontario and adjacent Quebec. Its blue flowers and dense stands make it one of the more spectacular plants of that area in late June and early July. It is rare or local elsewhere in Canada, although it has been found in every province.

Habitats   Rocky permanent pastures, abandoned fields, meadows, roadsides, and particularly on dry shallow soils over limestone.

Similar plants   Blueweed with its large blue flowers and rough hairy foliage, painful to the touch, is not likely to be mistaken for other plants.

Two other plants belonging to the same family as blueweed are often found with it on rocky pastures in limestone regions of Eastern Canada. Hound's-tongue, *Cynoglossum officinale* L., a biennial plant introduced from Europe, has soft hairy leaves, very large basal leaves up to 3 dm long, and dull red-purple flowers, which are succeeded by 4 large bristly seeds. The seeds are particularly troublesome in sheep-producing areas, as they become entangled in the wool. Gromwell, *Lithospermum officinale* L., a perennial plant from Europe, has slightly roughened leaves about 7.5 cm long, leaves with very prominent veins, small white flowers, and hard, shiny white or pale brown seeds resembling tiny pearls.

Blueweed, **Echium vulgare: A,** plant; **B,** flower; **C,** seed.

BORAGE FAMILY — BORAGINACEAE

BLUEBUR *Lappula echinata* Gilib.

Other names   Blueweed, burweed, stickseed, stickweed.

Description   Annual or winter annual with a mousy smell, spreading by seeds; stems rarely over 6 dm high, branching in the upper half, covered with stiff hairs that are usually flattened against the stem; leaves alternate, entire, sessile, narrow, 12-50 mm long, rough, hairy on both sides; flowers at first grouped at the top of the stem and later arranged on long slender branches; flowers small, petals blue, each flower giving rise to a group of 4 seeds; seeds about 2.5 mm long, brown at maturity, narrowed at one end, warty, one side with a central ridge, the other side with a double row of hooked prickles around the margin. Flowering in June and July, maturing seed from July.

Origin   Europe.

Distribution in Canada   Occurs in every province, far more abundant west than east.

Habitats   Grainfields, pastures, railway grades, roadsides, and waste places.

Notes   The prickly seeds stick to clothing and animal hair and are a source of trouble to sheep farmers.

Similar plants   Western bluebur, *Lappula redowskii* (Hornem.) Greene, is a native annual found in the western provinces on dry prairies, sandy places, railway grades, and roadsides. There is no evidence that this plant has spread east of Manitoba. Western bluebur differs from bluebur mainly in its seeds, which have only 1 marginal row of prickles (see illustration). Other differences are not well defined: western bluebur often branches from the base, is generally a shorter plant, and tends to be hairier and therefore somewhat grayer than bluebur.

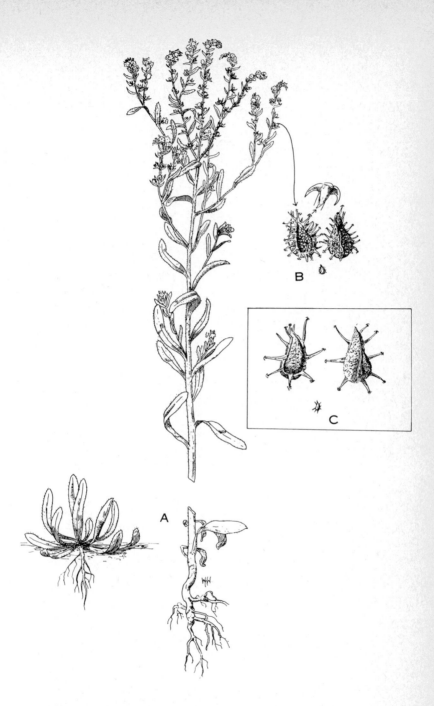

Bluebur, **Lappula echinata: A,** plant; **B,** outer and inner surfaces of seed and, above, a hooked prickle. Western bluebur, **Lappula redowskii: C,** outer and inner surfaces of seed showing a single row of prickles.

MINT FAMILY — LABIATAE

GROUND-IVY *Glechoma hederacea* L.

Other names   Creeping Charlie, gill-over-the-ground. Another scientific name, *Nepeta hederacea* (L.) Trev., has been applied to this plant.

Description   Perennial, spreading by seeds but mainly by creeping stems that root at the nodes; stems square; flowering stems 10-30 cm high, almost erect, unbranched; leaves opposite; leaf blades rarely over 25 mm wide, almost heart-shaped with regular rounded teeth, stalks slender; flowers from 1 to 7 in the leaf axils, petals united, forming a 2-lipped corolla, purple-blue, corollas 6-18 mm long; only the larger flowers with stamens and pistil, the others without stamens; calyx persistent, at maturity containing 4 seeds; seeds brown to dark brown, finely pebbled, about 1.5 mm long, rounded on one side and ridged on the other. Flowering from late April to the end of June.

Origin   Eurasia.

Distribution in Canada   Occurs in every province, but more common in Eastern Canada.

Habitats   Lawns, also in gardens and waste places. Sometimes planted in rock gardens where it soon becomes a pest.

Notes   There is some evidence that horses have been poisoned when they have eaten large quantities of ground-ivy. Other animals do not seem to be affected.

Similar plants   The leaf shape of ground-ivy is characteristic enough to permit easy separation from the speedwells, *Veronica* spp., also found in lawns. The speedwells, moreover, have rounded stems and flat heart-shaped capsules that contain many seeds.

Another member of the mint family, catnip, *Nepeta cataria* L., has leaves of similar shape to those of ground-ivy but is not found in lawns. Catnip is largely confined to roadsides, stony pastures, and areas close to farm buildings, particularly on lime-rich soils. The strong odor of bruised catnip, fascinating to cats, easily distinguishes catnip from ground-ivy. Catnip is a much taller plant than ground-ivy and its whitish flowers are arranged in a terminal inflorescence, not in the axils of well-developed leaves. Catnip is widely distributed in Canada, although only common in Ontario and Quebec. Catnip flowers from June to October.

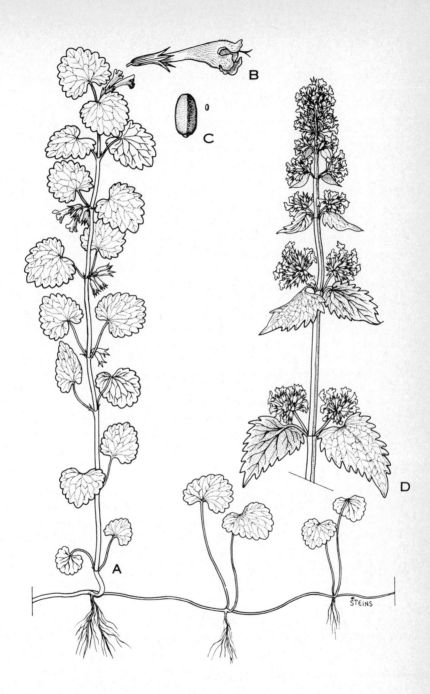

Ground-ivy, **Glechoma hederacea: A,** plant; **B,** flower; **C,** seed. Catnip,
Nepeta cataria: D, top portion of plant.

AMERICAN DRAGONHEAD *Dracocephalum parviflorum* Nutt.

Other names   Dragonhead, small-flowered dragonhead. Another scientific name, *Moldavica parviflora* (Nutt.) Britton, is sometimes applied to this plant.

Description   Annual or biennial with a strong taproot, spreading by seeds; stems 4-sided, erect, 3-9 dm high, often strongly branched, branches ascending; leaves opposite, stalked, coarsely toothed; flowers crowded into a dense terminal spike at the ends of branches or clustered in the axils of leaves, the spikes including spiny-tipped bracts as well as flowers; petals bluish or purplish, only slightly longer than the calyx; calyx with 15 strong nerves, 5-toothed, the uppermost tooth longer and broader than the others, calyx persistent and at maturity containing 4 seeds; seeds brownish black or black, not quite 3 mm long, rounded on one side, angled and ridged on the other, the angled side with a prominent slit towards the base. Flowering June to August, maturing seed from July until frost.

Origin   Native to North America. Introduced into Europe, possibly with wheat.

Distribution in Canada   More often found in the Prairie Provinces than elsewhere in Canada.

Habitats   Clover fields, grainfields, gardens, native grassland, clearings, edges of woods, and waste places.

Similar plants   Thyme-flowered dragonhead, *Dracocephalum thymiflorum* L., has been found only in Yukon Territory, Ontario, Manitoba, Saskatchewan, and Alberta, and rarely as a serious weed. A native plant of Asia, it is an annual or biennial and differs from American dragonhead as follows: less vigorous and usually shorter; flowers clustered in most of the leaf axils and not grouped in a terminal spike; bracts not spiny tipped; and seeds brown and only about 1.5 mm long.

Heal-all or self-heal, *Prunella vulgaris* L., has flowers clustered in spikes, but the perennial habit, leaves without teeth, and the conspicuous bracts of the spikes should serve for differentiation from American dragonhead.

The mint family to which the dragonheads belong is rather easily recognized by its aromatic odor, square stems, irregular flowers, opposite leaves, and clusters of 4 seeds produced in each flower.

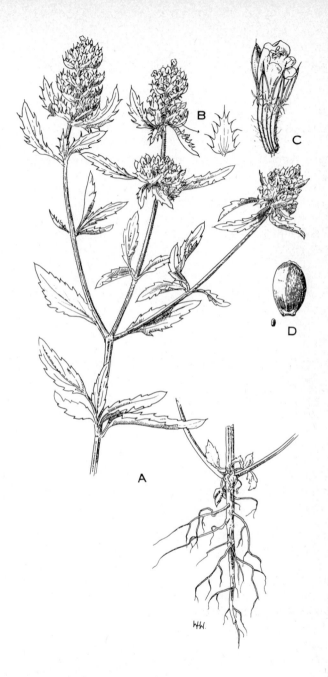

American dragonhead, **Dracocephalum parviflorum: A,** plant; **B,** bract;
C, flower; **D,** seed.

HEMP-NETTLE *Galeopsis tetrahit* L.

Other names   Dog, bee, or flowering nettle. Hemp-nettle is a very variable species in respect to flower color, flower size, and leaf shape. Plants with small, purplish flowers and leaves wedge-shaped at the base are sometimes called *Galeopsis tetrahit* L. var. *bifida* (Boenn.) Lej. & Court.

Description   Annual, spreading by seeds; stems 3-7 dm high, more or less branched, usually swollen below the joints, very stiff, bristly hairs are particularly numerous below the joints; leaves opposite, rather long stalked, wedge-shaped or rounded at the base, coarsely toothed except at the base, hairy, the long stiff hairs more abundant on the upper surface; flowers in dense clusters in the axils of the upper leaves, without stalks; petals (corolla) united, corolla 2-lipped, hairy, pale purple, white or variegated, 8 to nearly 25 mm long, longer than the calyx; calyx with 5 long, spiny-tipped lobes, hairy, at maturity surrounding a group of 4 seeds; seeds about 3 mm long, somewhat egg-shaped, the base with a prominent round scar, grayish brown with darker spots, the surface sprinkled with whitish warts. Flowering July to September.

Origin   Europe. Also Asia.

Distribution in Canada   Widely distributed in Canada in every province to the northern limits of agriculture. Although rare in the drier southern prairies, common in northern Alberta and a problem in the cultivated fields of the Neepawa-Minnedosa district and the Swan River valley in Manitoba.

Habitats   Grainfields, gardens, pastures, barnyards, waste places, and open woods.

Notes   Hemp-nettle appears to have originated as a hybrid between two European species of *Galeopsis*. When these two species were artificially crossed in Sweden, the resulting plants were indistinguishable from hemp-nettle.

Similar plants   Stems of hemp-nettle are covered with bristly hairs, which tend to penetrate the skin when the plant is handled. The groups of flowers are almost equally formidable, as the calyx is armed with 5 sharp points. The thickened stems below the joints are also characteristic. There is little possibility of mistaking hemp-nettle for other plants, even those of the mint family to which hemp-nettle belongs.

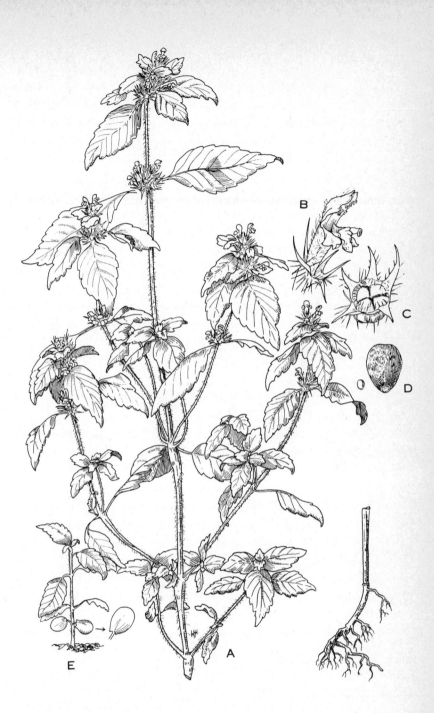

Hemp-nettle, **Galeopsis tetrahit: A,** plant; **B,** flower; **C,** mature flower with 4 seeds; **D,** seed; **E,** seedling.

FIGWORT FAMILY — SCROPHULARIACEAE

TOADFLAX *Linaria vulgaris* Mill.

Other names   Butter-and-eggs, wild snapdragon, yellow toadflax.

Description   Perennial, persistent by creeping roots; roots producing buds that develop into shoots; stems erect, little branched, glabrous or with short pubescence, 15-50 cm high, usually in clumps; leaves spreading in all directions, numerous, alternate, narrow, pointed, smooth, usually about 40 mm long but ranging from 25-75 mm; flowers borne along the upper part of the stem in long terminal clusters; petals (corolla) united, corolla 2-lipped, 25 mm long including the conspicuous basal spur, bright yellow with a deep orange center, rarely whitish; seedpods nearly 12 mm long, 2-chambered, each chamber containing many seeds; seeds dark brown or black, 2 mm in diameter including the broad flat margin (wing), circular in outline, strongly flattened, wing broad and with fine radiating ridges. Flowering throughout the summer.

Origin   Introduced from Eurasia probably as an ornamental. A native plant and weed throughout Europe and western Asia.

Distribution in Canada   Found in the agricultural areas of all provinces. Ranges as far north as Fairbanks, Fort Smith in the Northwest Territories, and Churchill, Manitoba. Very common in Eastern Canada. Introduced early to eastern North America. In 1758, in the writings of John Bartram, it was mentioned as a serious weed. In Western Canada, introduction has been much more recent, but rapid spread has resulted in serious infestations. Prolific production of strongly winged, readily dispersed seeds and efficient vegetative spread make the weed a potential menace to large areas of Western Canada.

Habitats   Waste places, roadsides, gardens, grasslands, and cultivated fields. Usually on lighter soils.

Notes   Before toadflax flowers, it is often mistaken for leafy spurge. A freshly broken stem or leaf will show a milky latex in leafy spurge but not in toadflax.

Similar plants   In recent years there have been a surprising number of reports, particularly from Saskatchewan, on the occurrence of Dalmatian toadflax, *Linaria dalmatica* (L.) Mill. This vigorous perennial species differs from toadflax in having larger flowers, about 4 cm long, broader leaves, and wingless seeds.

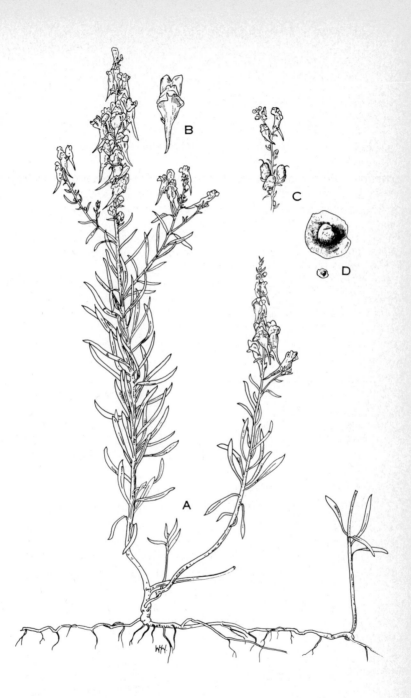

Toadflax, **Linaria vulgaris: A,** plant; **B,** flower; **C,** part of flowering stem with mature pods; **D,** seed.

PLANTAIN FAMILY — PLANTAGINACEAE

BROAD-LEAVED PLANTAIN *Plantago major* L.

Other names   Common plantain, dooryard plantain, whiteman's foot.

Description   Perennial, with a short thick rootstock fixed firmly in the soil by many tough strong roots, spreading by seeds; leaves in a basal rosette, dark green and dull, erect or prostrate, oval, entire or sometimes with a few coarse teeth, strongly ribbed, often slightly hairy, leaves on thick stalks that are usually about as long as the blades; flowers small and inconspicuous, greenish, grouped in narrow bluntish spikes on leafless stems; seedpods about 3 mm long, egg-shaped, brown, opening when ripe by splitting across near the middle to release the 5-16 seeds; seeds about 1 mm long, dark brown or nearly black, angular, variable in shape, finely marked with wavy threadlike ridges, one surface with a pale scar, seeds sticky when wet. Flowering throughout the summer.

Origin   Europe. Now distributed throughout most of the world.

Distribution in Canada   An abundant plant in the settled districts of all provinces.

Habitats   Pastures, meadows, cultivated land, gardens, lawns, roadsides, and waste places.

Notes   Pollens of the plantains are air-borne and are a factor in early summer hay fever at a time when the hay fever caused by grass pollens is also at its peak.

Similar plants   There are two other common weedy species of plantain in Canada. Rugel's or pale plantain, *Plantago rugelii* Dcne., is native to North America and occurs in Canada, largely in southern Ontario and southern Quebec, in much the same habitats as broad-leaved plantain. Rugel's plantain is similar to broad-leaved plantain, but differs in having paler green leaves, narrower spikes, narrower and longer seedpods opening near the base, and seeds roughened rather than ridged and longer than those of broad-leaved plantain. The most conspicuous difference lies in the leaf stalks, which are generally green near the base in broad-leaved plantain, and purplish red at the base in Rugel's.

Narrow-leaved or English plantain, *Plantago lanceolata* L., introduced from Europe, is found in Canada mainly east of the Great Lakes and in southern British Columbia. It differs from the other species as follows: narrower leaves; thicker and shorter flower spikes about 25 mm long; seeds 2.5 mm long, smooth and shiny, oblong, and hollowed on one side. Its seed is particularly hard to remove from red clover and alfalfa seed.

152

Broad-leaved plantain, **Plantago major: A,** plant; **B,** seed. Narrow-leaved plantain, **Plantago lanceolata: C,** plant without roots; **D,** seed.

COMPOSITE FAMILY *Compositae*

The composite family contributes more weedy species to the Canadian flora than any other family, and is represented on every roadside and in most cultivated fields, lawns, pastures, and hayfields by such common plants as dandelion, ox-eye daisy, sow-thistles, burdocks, thistles, goat's-beards, cockleburs, and ragweeds. Many of the most colorful ornamental and native plants belong to this family.

A brief description of the family is given here to provide an understanding of the flower parts essential for the identification of species. Some of the terms used in describing the composites are used only in descriptions of this distinctive family.

Particularly characteristic of the family is the grouping of small flowers into heads, which superficially resemble single large flowers. The head is surrounded by one or more rows of bracts, and this arrangement increases the resemblance of the head to a single flower with its calyx.

In this publication, the bracts are referred to as "bracts of the head." Other names used in technical works are "phyllaries," "tegules," or "involucral bracts." Bracts vary from species to species, in form, arrangement, texture, and number.

The stamens and styles are not mentioned in the descriptions of separate species, but are shown in the diagram on the opposite page. The stamens, 5 in number, are inserted on the petals and are united laterally, enclosing the style, which branches above into two arms.

The 5 petals are united into a corolla. The corolla may be tubular, and then has 5 short terminal lobes of equal length, or ligulate, and one side of the corolla tube prolonged and strap-shaped. Heads may contain ligulate flowers only, as in dandelion; or tubular flowers in the center and ligulate flowers at the margin, as in sunflower and ox-eye daisy; or tubular flowers only, as in ragweed. Plants with heads containing only ligulate flowers have milky juice in their leaves and stems. Plants with heads containing tubular flowers lack this milky juice.

The seeds in Compositae are, technically, achenes, that is, dry indehiscent fruits. For practical purposes "seed" is used in the descriptions of the species. The seeds are very often crowned by a pappus, a modified calyx, which may be of different form in individual species. It may consist of scales, bristles, simple or feathery (plumose) hairs in one or more rows, or may be absent. Illustrations of a number of seeds with attached pappus on the opposite page show some of the variations in seeds and pappus.

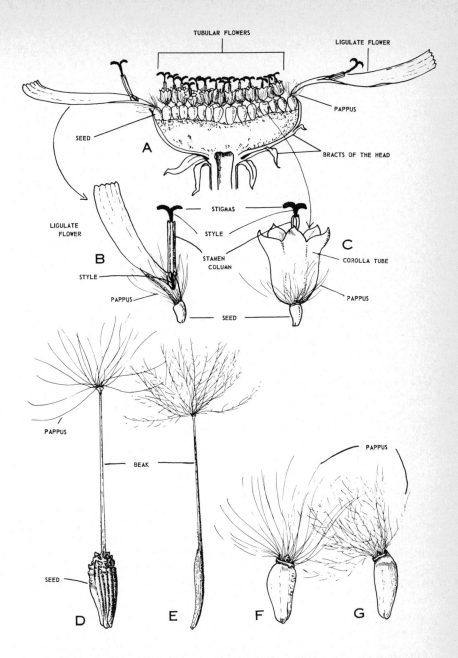

A, Cross section of flower head of a composite with tubular and ligulate flowers; **B,** ligulate flower; **C,** tubular flower. Seeds of several species showing some different types of pappus: **D,** simple pappus on a beaked seed—dandelion, **Taraxacum officinale; E,** plumose pappus on a beaked seed—spotted cat's-ear, **Hypochaeris radicata; F,** simple pappus—plumeless thistle, **Carduus acanthoides; G,** plumose pappus —Canada thistle, **Cirsium arvense.**

COMPOSITE FAMILY — COMPOSITAE

CANADA FLEABANE *Erigeron canadensis* L.

Other names   Bitterweed, fleabane, hog-weed, horse-weed, mare's-tail.

Description   Annual or winter annual, spreading by seeds; roots slender, short and often branched; stems erect, 7 cm to 18 dm high, usually simple up to the strongly branched inflorescence, generally bristly; leaves alternate, numerous, hairy, narrow, upper leaves or all leaves in smaller plants only about 5 mm broad, lower leaves in larger plants broader and with occasional teeth; flower heads very numerous, barely 4 mm broad, broader when seeds form, heads in a richly branched elongate and terminal inflorescence; bracts of the head narrow and pointed; flowers numerous and small; outer flowers ligulate and white, almost concealed by the bracts, tubular flowers of the center, yellow; pappus dirty white, almost three times as long as the seeds; seeds oblong, flattened, yellowish, hairy, 1.3 mm long. Flowering July to October.

Habitats   Cultivated fields, pastures, meadows, gardens, woodland trails, roadsides, and waste places. Usually dry sites, sometimes in sandy soils.

Origin   Native to North America. Introduced in many of the present localities.

Canada fleabane now has a practically worldwide distribution. It is considered to be the most completely naturalized plant of American origin in Europe.

Distribution in Canada   Common in the settled areas of all provinces. Less common in the provinces on the Atlantic Coast than inland.

Notes   The leaves cause a skin irritation in some individuals. Canada fleabane is reported to be irritating to the nostrils of horses.

Similar plants   Canada fleabane has ligulate flowers about as long as the bracts of the head, while the other common fleabanes have conspicuous ligulate flowers at least twice as long as the bracts.

The differences between asters and fleabanes are technical. The flower heads in asters are usually on leafy branches, whereas the heads of fleabanes are on naked stems. The ligulate flowers are narrower and more numerous in fleabanes and are usually in more than one row.

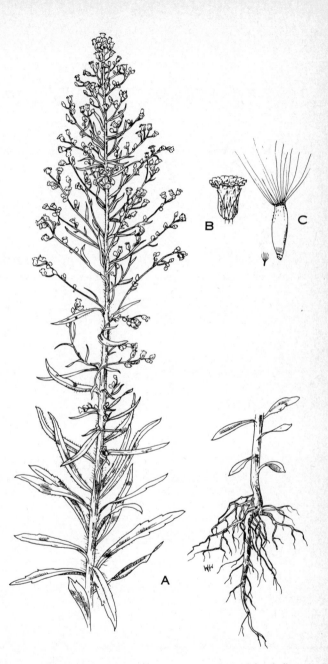

Canada fleabane, **Erigeron canadensis:** A, plant; B, head of flowers; C, seed with pappus.

COMPOSITE FAMILY — COMPOSITAE

PHILADELPHIA FLEABANE *Erigeron philadelphicus* L.

Description   Perennial by stolons and offsets; stems 3-9 dm high, soft hairy particularly at the base; leaves broad, alternate, lower leaves long-petioled and toothed, the upper sessile, clasping the stem, often entire; flower heads showy, drooping, nearly 25 mm broad; ligulate flowers very numerous, pale pink to white; seeds flattened, hairy. Flowering May to September.

Origin and distribution in Canada   Native to North America. Known from all provinces. Less common in the Prairie Provinces and the Maritime Provinces.

Habitats   Meadows, pastures, swampy ground, woods, riverbanks, beaches, and roadsides. Usually in moister habitats than the other weedy fleabanes.

Similar plants   Philadelphia fleabane is easily separated from the following two species by its clutching stem leaves.

ANNUAL FLEABANE *Erigeron annuus* (L.) Pers.

Description   Annual or biennial; stems erect, 15 cm to 15 dm high, usually with spreading hairs; leaves thin, edges conspicuously ciliate; basal and lower stem leaves narrowed into long petioles, coarsely toothed; middle and upper leaves narrower, sessile, less coarsely toothed, the uppermost often entire; heads 12 mm or more broad; ligulate flowers usually white, sometimes mauve-tinge.

Origin and distribution in Canada   Native to North America. Common in the eastern provinces. Rare in British Columbia and Manitoba.

Habitats   Meadows, pastures, cultivated land, gardens, waste places, and roadsides. A much more common plant in cultivated land than the following species.

Similar plants   Nearly all the stem leaves are toothed in the annual fleabane, and are entire or only the lower stem leaves toothed in the rough fleabane.

ROUGH FLEABANE *Erigeron strigosus* Muhl.

Description   Annual or biennial; stems erect, 6-12 dm high, often branched, appressed hairy; leaves alternate, firm, slightly roughened from appressed hairs, edges obscurely ciliate; basal leaves petioled, obovate, and commonly toothed; middle and upper leaves narrow, sessile, and entire; flower heads about 12 mm broad; ligulate flowers white or occasionally mauve.

Origin and distribution in Canada   Native to North America. Known from all provinces.

Habitats   A very common plant in old meadows and on roadsides, but rarely found in cultivated land.

Philadelphia fleabane, **Erigeron philadelphicus: A,** plant; **B,** seed with pappus. Annual fleabane, **Erigeron annuus: C,** leaf.

COMPOSITE FAMILY — COMPOSITAE

CANADA GOLDENROD *Solidago canadensis* L.

Description   Perennial, spreading by seeds and rootstocks; stem 4-15 dm high, hairy below the lower branches of the inflorescence; leaves numerous, 3-veined, glabrous above and hairy on the veins beneath or roughly hairy on both surfaces, most of the leaves except the uppermost with sharp teeth along the margins; flower heads 2.5 mm broad and about 3 mm long, on short stalks, numerous, crowded along one side of the inflorescence branches and forming a pyramidal panicle; bracts of the head in 3 or 4 rows, yellowish green; flowers minute, ligulate on the margin, and tubular in the center, both types of flower yellow; seeds hairy, about 0.8 mm long; pappus of simple hairs, soon falling. Flowering from late July.

Origin   Most of the approximately 120 known species occur in North America. Only one species is native to Europe and Asia.

Distribution in Canada   From Newfoundland to Saskatchewan. Goldenrods as a group occur throughout the cultivated and nonwooded areas across Canada.

Habitats   Colorful plants of meadows, old fields, roadsides, fencerows, and waste places. Goldenrods do not persist under cultivation.

Notes   Contrary to the opinion often held, goldenrods play a very unimportant part in hay fever. The flowering period of these plants coincides with the season of greatest suffering from hay fever and, as these are conspicuous plants, they are often suspected. The pollen of goldenrods can certainly produce hay fever symptoms, but usually the heavy sticky pollen is carried by insects or it drops to the ground close to the plant. Only occasionally, in dry, very windy weather enough goldenrod pollen is blown into the air to disturb sensitive people.

Canada goldenrod and a few other goldenrods are much beloved garden plants in Europe.

Similar plants   Canada goldenrod is one of a large group of plants known as goldenrods. All goldenrods are perennial plants with rootstocks, small flower heads, leafy stems, sessile or nearly sessile stem leaves, and yellow flowers (cream color in *Solidago bicolor* L.). Separation of the species is based on technical characters and is not attempted here.

Canada goldenrod, **Solidago canadensis: A,** plant; **B,** seed with pappus.

COMPOSITE FAMILY — COMPOSITAE

POVERTYWEED *Iva axillaris* Pursh

Other name   Small-flowered marsh elder.

Description   A persistent perennial with an unpleasant odor, spreading by seeds and rootstocks; stems erect, 15-45 cm high, branched, in dense clumps; leaves opposite in the lower part of the stem, alternate above, oblong, obtuse, thick, entire, rough hairy, sessile, pale green, usually less than 25 mm long; flower heads greenish, solitary at bases of the upper leaves, drooping, small, less than 5 mm broad, all heads contain male and female flowers, the few marginal flowers are female and produce seeds, the other flowers are male; bracts 5, united into a cup; seeds grayish to black, dull, plump, wedge-shaped, up to 5 mm long, often with resinous dots; pappus lacking. Flowering June to August.

Origin   Western North America. Reported from Australia as an introduction.

Distribution in Canada   Occurs in all the Prairie Provinces and apparently of greatest abundance in Saskatchewan. Less common in British Columbia and not yet introduced into the eastern provinces. Found as far north as the Peace River.

Habitats   Cultivated land, eroded hillsides, poorly drained clay flats, roadside ditches, waste places, and saline and alkaline flats and meadows. Frequently in heavy alkaline soils of poor physical condition.

Notes   Povertyweed is a wind-pollinated plant and its pollen produces hay fever in sensitive individuals. Single plants shed little pollen, but when povertyweed is abundant it may be an important cause of hay fever.

Similar plants   Povertyweed is not likely to be mistaken for other plants. Perennial ragweed also has creeping rootstocks and is about the same height as povertyweed, but perennial ragweed has lobed leaves and the flower heads contain only one kind of flower.

162

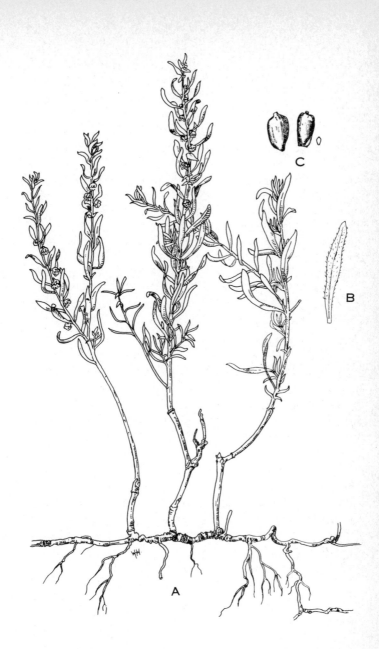

Povertyweed, **Iva axillaris: A,** plant; **B,** leaf showing stiff hairs; **C,** seeds.

COMPOSITE FAMILY — COMPOSITAE

FALSE RAGWEED *Iva xanthifolia* Nutt.

Other names   Burweed marsh elder, careless weed, marsh elder, prairie rag-weed.

Description   Annual, spreading by seeds; taproot slender; stems 9-24 dm high, glabrous; leaves mostly opposite with long, hairy stalks, large, coarsely and irregularly toothed, both surfaces with appressed hairs, velvety; heads small, about 5 mm broad, greenish, in leafless loose branching clusters at the top of the plant and in the axils of the upper leaves; the heads contain male and female flowers, the marginal flowers are female and produce seeds, the central flowers are male and produce pollen; bracts of the head 5, separate, acute; seeds brownish to black, wedge-shaped, to 3 mm long, finely striate, each flower head usually with 5 seeds; pappus lacking. Flowering mainly in late August and early September.

Origin   Native in the western prairies. Introduced into the eastern provinces and Northeastern United States. False ragweed has been introduced into Great Britain and Continental Europe.

Distribution in Canada   Abundant in Manitoba, Saskatchewan, and Alberta. Much less common in British Columbia, Ontario, and Quebec.

Habitats   Particularly common around towns and on roadsides. Also found in cultivated land, waste land, along railroads, and in gardens.

Notes   Contact with the leaves produces a dermatitis in some people. Milk from cows grazing on the leaves has an undesirable flavor. The abundant pollen is an important cause of hay fever.

Similar plants   Before it flowers, false ragweed may be mistaken for a number of plants: cocklebur, *Xanthium* spp., giant ragweed, *Ambrosia trifida*, and sunflowers, *Helianthus* spp. The leaves and stems of these plants are rough to touch, while the stems of false ragweeds are smooth and the leaves are velvety. Also, the leaves of giant ragweed are generally deeply cut into three lobes and the leaves of cocklebur are alternate.

Flower heads of false ragweed contain male and female flowers, whereas in ragweeds, *Ambrosia* spp., the heads contain either male or female flowers, never both.

False ragweed, **Iva xanthifolia: A,** plant; **B,** head of flowers showing the male flowers surrounded by the 5 pistillate flowers; **C,** seeds.

COMPOSITE FAMILY — COMPOSITAE

COMMON RAGWEED *Ambrosia artemisiifolia* L.
var. *elatior* (L.) Descourtils

Other names   Roman wormwood, short ragweed, small ragweed. Formerly known by the scientific name *Ambrosia elatior* L.

Description   Annual, spreading by seeds; stems erect, usually 6-9 dm high, bushy branched, rough hairy; leaves opposite or alternate, short-stalked, thin, usually twice divided into narrow segments; flower heads contain either male or female flowers; the 1-flowered female heads are sessile and inconspicuous, and borne singly or in small groups in the axils of the upper leaves, plants rarely bear only female heads; the male heads are more numerous and are in spikes terminating the stems and branches; bracts of the male heads united, umbrellalike; "seeds" less than 4 mm long, consisting of a brittle, woody outer covering with one central terminal beak surrounded by a ring of sharp teeth, within this outer covering is the whitish, oily, soft true seed enclosed in a smooth brown coat; pappus lacking. Flowering mainly in August and early September.

Origin   Native to North America.

Distribution in Canada   Occurs in every province. Far commoner in southern Ontario and southern Quebec, east as far as Quebec City, than elsewhere in Canada. Very rare in British Columbia and Newfoundland, uncommon in the Prairie Provinces, and not, as yet, a serious threat in the greater part of the provinces on the Atlantic Coast.

Habitats   Common ragweed is found under a wide variety of soil and moisture conditions in cultivated fields, gardens, vacant lots, waste places, along roadsides, and fencerows. A typical after-harvest cover in grainfields and hayfields.

Notes   This species is the most abundant of the ragweeds and the most important cause of hay fever in eastern North America. The plant or its pollen may produce a dermatitis in some people who are not necessarily sufferers from hay fever. Dairy products from cows that have grazed on this plant have an objectionable odor and taste.

Similar plants   The perennial ragweed, *Ambrosia psilostachya* DC. var. *coronopifolia* (T. & G.) Farw., has horizontal perennial rootstocks (*D* in illustration). Apart from this character, it resembles the common ragweed in appearance, although it is usually a smaller plant with rougher, thicker, and less-lobed leaves. The perennial ragweed is a common native in Western Canada and in recent years has spread eastward along railway lines.

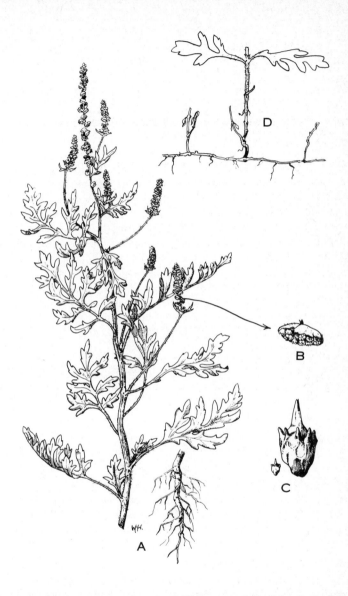

Common ragweed, **Ambrosia artemisiifolia** var. **elatior: A,** plant; **B,** head of male flowers; **C,** "seed." Perennial ragweed, **Ambrosia psilostachya** var. **coronopifolia: D,** lower part of plant showing underground root-stocks.

COMPOSITE FAMILY — COMPOSITAE

GIANT RAGWEED *Ambrosia trifida* L.

Other names   Great ragweed, kinghead, tall ragweed.

Description   Annual, spreading by seeds; stems less than 30 cm high to over 3 m, rough hairy; leaves opposite, hairy, long-petioled, margins saw-toothed, prominently 3-lobed, unlobed, or 5-lobed; lobed and unlobed leaves often occur on the same plant and sometimes all leaves may be without lobes; flowers in greenish heads, each head containing either female (seed-producing) flowers or male (pollen-producing) flowers, never both; the 1-flowered seed-producing heads are clustered in groups of 1-3 at the bases of the male spikes or in the axils of the upper leaves; the male heads are much more numerous and are in spikes terminating the stems and branches; bracts of the male heads united and with 3 strong black ribs on one side; "seeds" 6-9 mm long, consisting of a thick, woody, more or less mottled outer covering, which at the upper end has 1 central beak surrounded by a circle of 5 or more marginal points, within this outer covering is the whitish, oily, soft, true seed enclosed in a smooth black coat; pappus lacking. Flowering mainly from late July to early September, usually in flower a few days before common ragweed.

Origin   Native to North America.

Distribution in Canada   Known from all provinces except Newfoundland. Rare in British Columbia and Alberta, not very common in Saskatchewan. The most abundant ragweed in Manitoba. Common in southern Ontario and south-western Quebec, but far less abundant than common ragweed and of much less importance as a hay fever plant.

Habitats   Found along roadsides, railways, in agricultural fields, and in waste places near towns, usually on rather rich soils. It is sometimes found in more undisturbed habitats, in marshes that dry out in summer, or on rich moist soils near streams, and reaches its greatest growth under these conditions.

Notes   Wind-blown pollen from this plant is a common cause of autumn hay fever.

Similar plants   False ragweed (page 164) differs from this plant in having smooth stems and leaves without definite lobes.

Giant ragweed, **Ambrosia trifida: A,** plant; **B,** head of male flowers from above showing 3 distinct dark lines; **C,** head of male flowers in side view; **D,** "seed."

HAIRY GALINSOGA *Galinsoga ciliata* (Raf.) Blake

Other names   Ciliate galinsoga, quick-weed.

Description   Annual, spreading by seeds; stems erect, 15-60 cm high, often much branched above, with scattered to abundant, spreading, often glandular hairs; leaves opposite, stalked, broadly oval, and tapering to a point at the apex, 25-60 mm long, margins coarsely toothed, scattered hairs on both surfaces; heads small, about 6-8 mm across, on long stalks from the axils of upper leaves; bracts of the head green, few; flowers of two kinds, about 5 white ligulate flowers around the margins and more numerous yellow tubular flowers in the center; seeds dark brown to black, oblong, about 1.5 mm long, usually with minute hairs; seeds of both ligulate and tubular flowers with a short pappus of separate white scales about 0.8 mm long. Flowering from mid-June to October.

Origin   Introduced from tropical South America. Shows its origin by being one of the first plants to succumb to frost.

Distribution in Canada   Found in all provinces except Newfoundland. Most abundant in Quebec, Ontario, and parts of British Columbia.

Habitats   Usually in gardens and other habitats about settlements. Also a weed of cultivated fields in the Fraser Valley of British Columbia.

Similar plants   Small-flowered galinsoga, *Galinsoga parviflora* Cav., closely resembles hairy galinsoga. Its flowers are the same size as those of hairy galinsoga, not smaller as is suggested by the common name. Small-flowered galinsoga differs from hairy galinsoga in having its stem glabrous or nearly so and in the absence of a pappus on the seeds of the ligulate flowers. Small-flowered galinsoga has also been introduced from tropical America, but in Canada is known only from single localities in Quebec, Ontario, and Manitoba.

Both species have been introduced to the Old World, but small-flowered galinsoga is by far the commoner species there.

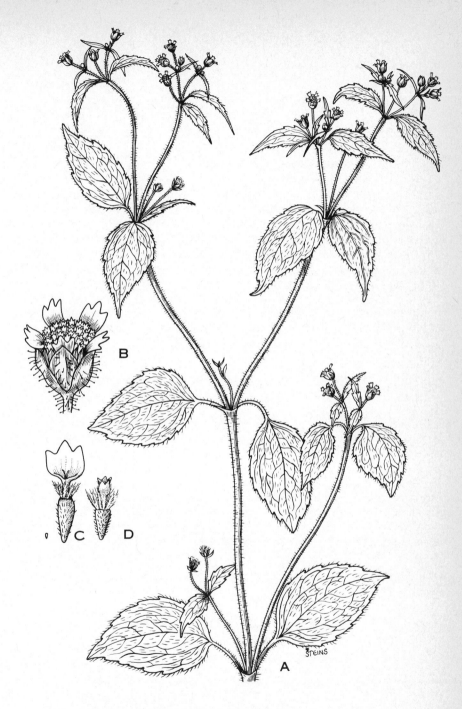

Hairy galinsoga, **Galinsoga ciliata: A,** plant; **B,** head of flowers; **C,** seed of ligulate flower with pappus; **D,** seed of tubular flower with pappus.

YARROW *Achillea millefolium* L.

Other names   Common yarrow, milfoil. *Achillea lanulosa* Nutt. is sometimes applied to yarrow that is native to inland North America, and *Achillea borealis* Bong. to yarrow that is native to the north and along both seacoasts.

Description   Perennial, spreading by seeds and shallow, horizontal rootstocks; flowering stems erect, to 6 dm high, woolly or glabrous; leaves in basal rosettes and alternate along the stem, finely cut into narrow lobes, which are also divided, hairy; flower heads small, about 6 mm across when in flower, crowded in dense, flat-topped clusters; bracts of the head in several overlapping rows, hairy, edges of bracts most often brown, sometimes black, and occasionally transparent; flowers of two kinds; ligulate flowers short and broad, white, rarely pink or purplish, usually 5 ligulate flowers and thus the head resembles a simple flower, and tubular flowers, white, numerous; seeds oblong, about 1.5 mm long, finely lined lengthwise, thick margined; pappus lacking. Flowering throughout the summer.

Origin   It was formerly thought that most of the yarrow found from southern Newfoundland to Ontario was introduced from Europe. However, recent investigations have shown that almost all of the yarrow in Canada is native to North America. Only a form with deep-purple flowers, an escape from gardens, is certainly of European origin.

Distribution in Canada   Newfoundland to British Columbia. With the exception of the dandelion, yarrow is perhaps the most commonly occurring weed in Canada.

Habitats   Pastures, lawns, meadows, roadsides, and waste places.

Notes   Yarrow has an aromatic odor and bitter taste. Dairy products from cows grazing on this plant may have an undesirable flavor.

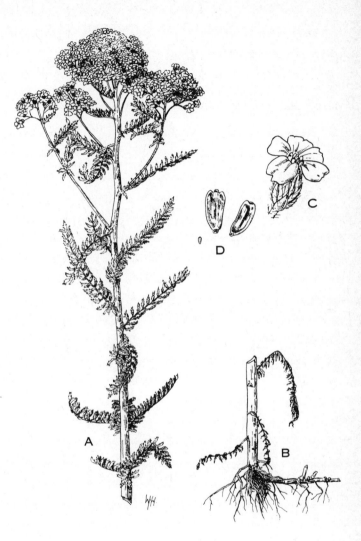

Yarrow, **Achillea millefolium: A,** plant; **B,** lower part of plant with root-stock; **C,** head of flowers; **D,** seeds.

COMPOSITE FAMILY — COMPOSITAE

SCENTLESS CHAMOMILE *Matricaria maritima* L.
var. *agrestis* (Knaf) Wilmott

Other names   False chamomile. Sometimes known by the scientific name *Matricaria inodora* L.

Description   Annual to short-lived perennial, flowering in the first year from seed; stems erect or semierect, glabrous, often much branched, 15 cm to 1 m high; leaves alternate, mostly sessile, finely dissected into narrow segments, usually glabrous; flower heads solitary at the ends of long branches, showy, 18-30 mm in diameter; flowers of two kinds: a single row to occasionally several rows of white ligulate flowers around the margins, and numerous yellow tubular flowers in the center; the swollen portion of the flower head on which the flowers are attached (receptacle) is naked (lacks scales); bracts of the head numerous, in several overlapping rows; seeds about 2 mm long, dark brown, one side with 3 broad ribs and 2 intervening dark-brown grooves and the other with a broad central dark-brown area running from the base to the apex where it terminates in 2 dark-brown oil glands; pappus minute or lacking. Flowering late June to September.

Origin   Europe.

Distribution in Canada   In all provinces. Most common in the Maritime Provinces and the adjacent part of Quebec and in the Prairie Provinces. Although eastern and western plants are identical in appearance, eastern plants have half as many chromosomes in their cells as have the western plants.

Habitats   Cultivated fields, roadsides, and waste places. A serious weed of cultivated fields only in the Prairie Provinces.

Similar plants   Wild chamomile, *Matricaria chamomilla* L., stinking mayweed, *Anthemis cotula* L., and corn chamomile, *Anthemis arvensis* L., introduced from Europe, have flowers and leaves similar to those of scentless chamomile. Crushed leaves of wild chamomile and stinking mayweed have a strong odor, whereas scentless chamomile leaves are practically odorless. Stinking mayweed and corn chamomile, *Anthemis* spp., have stems that are hairy below the heads, and receptacles with scales. Scentless and wild chamomile, *Matricaria* spp., have glabrous stems and lack scales on the receptacles.

The most common of these similar plants is stinking mayweed. It is found around farm buildings, roadsides, waste places, and occasionally in cultivated fields, in Prince Edward Island, Nova Scotia, New Brunswick, Ontario, and British Columbia. It is a rare plant in the Prairie Provinces.

Pineapple weed, *Matricaria matricarioides* (Less.) Porter, is closely related to scentless chamomile, but can easily be distinguished from it and similar plants. Pineapple weed is a much smaller plant, lacks white ligulate flowers, and its crushed leaves produce a distinct pineapple odor. It is found in all provinces and around settlements in Yukon Territory and Northwest Territories. Pineapple weed usually occurs around buildings, gardens, roadsides, and waste places, particularly on trampled ground.

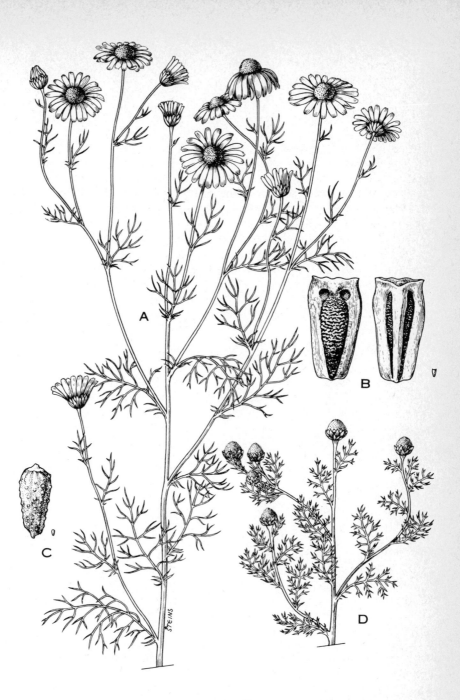

Scentless chamomile, **Matricaria maritima** var. **agrestis: A,** plant; **B,** seed.
Stinking mayweed, **Anthemis cotula: C,** seed. Pineapple weed, **Matricaria matricarioides: D,** plant.

OX-EYE DAISY *Chrysanthemum leucanthemum* L.

Other names   Marguerite, white daisy, white-weed.

Description   Perennial, spreading by seeds and slowly extending shallow root-stocks; stems 3-9 dm high, simple or branched near the top, one to several stems from the same root, smooth or sparingly hairy at the base; stem leaves alternate, mostly sessile and somewhat clasping, deeply cut in most plants, usually glabrous; basal and lowest leaves petioled, broader, and less deeply cut than the stem leaves; flowering heads solitary at the end of long stalks, showy, 30-50 mm broad; tubular flowers of the center numerous, yellow; the ligulate flowers at the margin of the head 15-35 in number, white, 12 to nearly 25 mm in length, with 2 or 3 shallow teeth at the tip; bracts of the head in several overlapping rows, numerous, green with brown margins; seeds 1.7-2.5 mm long, black with usually 10 prominent, rounded, white ribs; pappus lacking. Flowering through-out the summer, more abundantly in July.

Origin   Europe. Introduced at a very early date.

Distribution in Canada   Of greatest abundance in Eastern Canada, particular-ly in the Atlantic Provinces. Rare in Saskatchewan and most of Alberta. Wide-ly distributed in the Peace River district of Alberta and in British Columbia. Collections have been made in Labrador and at the northernmost tip of New-foundland.

Habitats   Old meadows and pastures, along roadsides and railways, and lawns. A major weed of the grasslands of northeastern America.

Notes   When eaten by dairy cattle, ox-eye daisy gives milk a disagreeable taste. In the first year from seed, ox-eye daisy does not develop blooms and the nonflowering leafy rosettes may easily be overlooked.

Similar plants   A less common form of ox-eye daisy with leaves not so deeply cut occurs in the provinces on the Atlantic Coast, but for all practical purposes it need not be differentiated.

The scentless chamomile, *Matricaria maritima* L. var. *agrestis* (Knaf) Wilmott, and stinking mayweed, *Anthemis cotula* L. (page 174), also have white ligulate flowers and yellow tubular flowers. Both plants differ from ox-eye daisy in being annuals and in having smaller flowering heads, and much more finely dissected leaves.

Ox-eye daisy, **Chrysanthemum leucanthemum: A,** plant; **B,** seed.

ABSINTHE *Artemisia absinthium* L.

Other names   Absinthium, wormwood.

Description   Strongly aromatic perennial, forming a rosette in the first year, spreading by seeds; stems to 15 dm high, several from each rosette, silver-gray and covered with fine hairs, grooved, woody at the base; leaves alternate, covered with fine hairs, greenish above and grayish below, divided into narrow and rather blunt segments, the lower leaves with long stalks, the upper leaves almost stalkless; flower heads small, about 4 mm across, on stalks equaling to slightly longer than the heads, densely crowded in the axils of small leaves on the upper branches; flower heads containing only tubular flowers, flowers yellow and numerous; the swollen portion of the flower head on which the flowers are attached (receptacle) densely covered with long hairs; seeds very small, about 1 mm long, broader at the tip than the base, finely streaked, brownish. Flowering from late July to September.

Distribution in Canada   Found in all provinces. Particularly abundant in the Prairie Provinces, although not recognized as a serious weed until 1954.

Habitats   Roadsides, waste places, farmyards, pastures, and cropland.

Notes   When eaten by cattle, absinthe causes taint in dairy products. Grain has been rejected because a small amount of absinthe was present. Absinthe is used in the preparation of absinthe, vermouth, and other alcoholic beverages and was formerly valued for medicinal purposes. Its volatile oils are toxic if consumed in large amounts.

Similar plants   Absinthe resembles some of the native and introduced wormwoods, *Artemisia* spp. Its strong aromatic odor, tall coarse stems, rosette leaves with long stalks, and divided and silky hairy leaves that are only slightly darker on the upper surface should be helpful in identification.

Absinthe, **Artemisia absinthium: A,** plant; **B,** seed.

COMPOSITE FAMILY — COMPOSITAE

TANSY RAGWORT *Senecio jacobaea* L.

Other names   Common ragwort, staggerwort, stinking Willie.

Description   Biennial and perhaps occasionally a short-lived perennial, forming rosettes in the first year, spreading by seed; stems erect, 3-9 dm high, usually glabrous, branched above the middle; leaves alternate, dark green, deeply cut into irregular segments; rosette and basal leaves stalked; stem leaves without stalks, embracing the stem; flower heads numerous, in flat-topped clusters; bracts of the head in a single row; ligulate flowers yellow, conspicuous, and about 6 mm long; tubular flowers yellow; seeds of the ligulate flowers glabrous, those of the tubular flowers densely hairy, about 1.5 mm long, prominently ribbed; pappus of soft white hairs about 3 times as long as the seed. Flowering in late July and August.

Origin   Europe.

Distribution in Canada   Found in all the provinces on the Atlantic seaboard and in British Columbia, where it occurs on Vancouver Island and in the Fraser Valley. Also near Guelph, Ontario, in Puslinch Township.

Habitats   Pastures, hayfields, roadsides, and waste places.

Notes   Tansy ragwort contains a toxic alkaloid that has caused considerable loss of cattle and horses in various parts of the world. At the beginning of the century, the average yearly toll of cattle in Pictou County in Nova Scotia was 200 head. Losses from this cause are now almost unknown.

Similar plants   Tansy, *Tanacetum vulgare* L., has yellow flowers in dense flat-topped clusters and deeply cut leaves and thus to some extent resembles tansy ragwort. Tansy may be distinguished from tansy ragwort in that it lacks ligulate flowers, the pappus forms a very short crown, and the leaf segments are regularly and sharply toothed.

Many species of ragwort occur in Canada. Tansy ragwort may be readily distinguished from other species by the leaf shape. One western native plant, cut-leaved ragwort, *Senecio eremophilus* Richards, rather closely resembles tansy ragwort, but the leaves are not so freely lobed, and they taper to a long narrow, acute point (in tansy ragwort the leaf is rounded in outline at the apex), and all the seeds are without hairs.

Common groundsel, *Senecio vulgaris* L., is a slender annual or winter annual plant differing from other annual species of *Senecio* and tansy ragwort in having black-tipped bracts and inconspicuous flower heads without ligulate flowers. Common groundsel, an introduction from Europe, is found in all provinces, mostly in rich, cultivated soil.

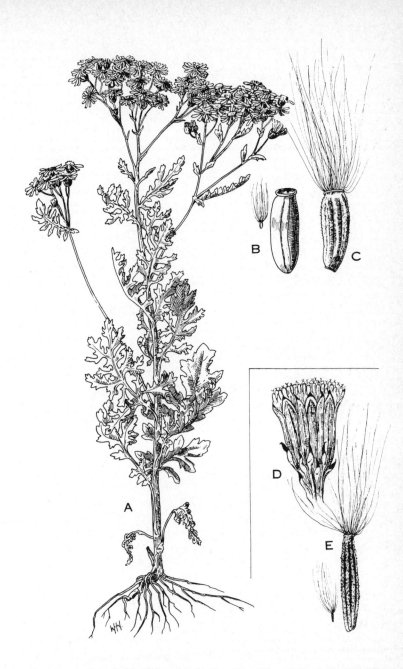

Tansy ragwort, **Senecio jacobaea: A,** plant; **B,** seed of ligulate flower
with pappus removed; **C,** seed of tubular flower. Common groundsel,
Senecio vulgaris: D, head of flowers; **E,** seed.

COMMON BURDOCK *Arctium minus* (Hill) Bernh.

Other names   Bardane, clotbur, lesser burdock, wild burdock.

Description   Biennial, spreading by seeds; stem 6-18 dm high, erect, grooved, hollow, branched, rough hairy; taproot long, thick, and fleshy; leaves alternate, large, those at the base often 3 dm broad, more or less heart-shaped, woolly hairy on the lower surface at first but eventually glabrous or nearly so, practically smooth above, stalks of the lower leaves stout and hollow; flower heads numerous, on short stalks or stalkless, clustered at the ends of branches and along the stem in the axils of leaves, less than 25 mm across; bracts of the head numerous, glabrous or slightly woolly, inner bracts shorter than the flowers, all bracts extending into hooked spines; flowers usually purple, rarely white, all alike and tubular, protruding from the envelope of bracts; pappus of barbed bristles, about 3 mm long, soon falling; seeds oblong, 4-6 mm long, mottled, somewhat ridged. Flowering from mid-July to mid-September.

Origin   Europe.

Distribution in Canada   Known from every province. Most abundant in Eastern Canada. The other burdocks are far less common, but are sometimes locally abundant.

Habitats   Farm yards, fencerows, roadsides, waste places, and stream banks. Usually on moist fertile soils. Rarely troublesome on cultivated land or in pastures.

Notes   The mature flower head forms a prickly bur readily distributed on clothing or by animals.

Similar plants   Burdock forms a rosette of large leaves in the first year from seed. At this stage it somewhat resembles rhubarb; however, the whitish undersurface of the burdock leaves permits easy differentiation.

When mature, the plant is not likely to be mistaken, except for other species of burdock. Great burdock, *Arctium lappa* L., differs from common burdock in the following ways: heads larger, over 25 mm broad; flower heads widely open in fruit; the bracts of the head are all about the same length and equal the flowers; the stalks supporting the heads are longer; the heads are arranged in a flattish topped inflorescence, not scattered along the stems; the stalks of the lower leaves are solid not hollow. Woolly burdock, *Arctium tomentosum* Mill., differs from the common burdock in having very cottony heads that are on longer stalks and form a flat-topped inflorescence.

Common burdock, **Arctium minus: A,** plant; **B,** rosette; **C,** seeds.

COMPOSITE FAMILY — COMPOSITAE

NODDING THISTLE *Carduus nutans* L.

Other name   Musk thistle.

Description   Biennial, forming a large, flat rosette the first year and flowering the second year, spreading by seed; taproot long and fleshly; stems spiny-winged except just below flower heads, 3-18 dm high, rarely with more than a few branches; leaves alternate, deeply lobed with spiny teeth; bracts of the flower head hairy to glabrous, 3-6 mm wide, with strong sharp prickles at the tip, outer bracts bent at right angles to the flower head (reflexed); flower heads few to many, nodding, 25-60 mm in diameter; flowers purple, all tubular; seeds nearly 4 mm long, light brown; pappus long, simple, white. Flowering from July to September.

Origin   Eurasia.

Distribution in Canada   Known from every province except Prince Edward Island and Alberta. Most common in Ontario and Saskatchewan.

Habitats   Pastures, roadsides, and waste places. In Eastern Canada it often forms very dense stands on rocky, hilly soil, where there is little competition.

Notes   In Saskatchewan these plants have larger, glabrous bracts, whereas plants from elsewhere usually have smaller bracts with weblike hairs. The plants in Saskatchewan are thought to have an independent origin, possibly from Argentina in early importations of rapeseed.

Similar plants   Nodding thistle is closely related to the plumeless thistle, *Carduus acanthoides* L., with which it hybridizes in some localities in Ontario. Nodding thistle can be distinguished from all other thistles, including the plumeless thistle, by its nodding flower head, reflexed bracts, and naked stem below the head.

Plumeless thistle most closely resembles Canada thistle, *Cirsium arvense* (L.) Scop., in appearance, but differs from it in having green, prickly wings on the stem, perfect flowers, a pappus that lacks side branches, and a biennial habit of growth. Plumeless thistle is a native of Europe and is found in Nova Scotia, Quebec, Ontario, and British Columbia. It is most common on rocky hilly pastures, roadsides, and waste places in Ontario.

Nodding thistle, **Carduus nutans: A,** plant; **B,** seed with pappus.

CANADA THISTLE *Cirsium arvense* (L.) Scop.

Other names   Creeping or field thistle. Popular usage has established Canada thistle as the common name for this plant, although it is actually of European origin. Creeping thistle, the name given to the plant in England, is far more descriptive and exact.

Description   Perennial, spreading by seeds and whitish roots; stems erect and green, to 12 dm high, branched; leaves alternate, irregularly lobed, spiny-toothed; flower heads numerous, small compared with those of other thistles, globular in the male plants, more or less flask-shaped in the female; bracts of the head ending in weak prickles; flowers variable in color, rose-purple, pink or less often white, all tubular; flowers either male or female, each sex on separate plants, seeds nearly 3 mm long, oblong, flattened, somewhat curved, smooth, dark brown; pappus feathery, copious, and white, easily separating from the seed. Flowering throughout the summer.

Origin   Europe, Western Asia, and Northern Africa. Probably introduced from Europe.

Distribution in Canada   Throughout the agricultural areas of all provinces.

Habitats   The commonest and most troublesome thistle in cultivated fields, meadows, pastures, roadsides, and waste places. Grows under a wide variety of soil and moisture conditions, although less common or absent on light dry soils.

Notes   Large patches of Canada thistle may consist entirely of male plants. If such patches are examined no seed will be found, and this, in part, accounts for the opinion that this thistle is a poor seed producer. Patches of female plants not too far removed from male plants produce abundant seed, although often immature seeds are destroyed by larvae of various insects.

Similar plants   A number of varieties of Canada thistle occur rather rarely. These varieties are less spiny than the thistle illustrated, and one of them is distinctly woolly on the undersurface.

Canada thistle is the only thistle with male and female flowers on separate plants. It is also readily distinguished from other thistles by the combination of green stems without spiny wings, small almost spineless heads, and creeping roots.

The plumeless thistles, *Carduus* spp. (page 184), have winged stems and the pappus is not feathery. All true thistles, *Cirsium* spp., have a feathery pappus.

Canada thistle, **Cirsium arvense: A,** plant; **B,** seed with pappus; **C,** seed with pappus removed.

COMPOSITE FAMILY — COMPOSITAE

BULL THISTLE *Cirsium vulgare* (Savi) Tenore

Other names   Spear thistle. Another scientific name formerly used was *Cirsium lanceolatum* Hill.

Description   Biennial, forming a large flat rosette in the first year, spreading by seed; taproot deep and fleshy; stems spiny-winged and slightly woolly, 3-15 dm high, branched above; leaves alternate, deeply lobed, the lobes ending in a very long sharp prickle with shorter prickles between them, upper surface of leaf with short prickles, lower surface white hairy; bracts of the head somewhat hairy, tipped with long sharp prickles, prickles usually yellowish at the tip; flower heads large, 40-75 mm across when mature; flowers purple, all tubular; seeds nearly 4 mm long, smooth or slightly furrowed lengthwise, blackish or black-streaked on a yellowish background; pappus long, feathery, white. Flowering throughout the summer.

Origin   Europe and Asia.

Distribution in Canada   A very common plant in Eastern Canada and southern British Columbia. It is becoming abundant around Lethbridge in southwestern Alberta and has been collected at the Cypress Hills in Alberta and Saskatchewan.

Habitats   Pastures, old fields, roadsides, and waste places. Usually on rich, moist soils. These habitats indicate that bull thistle, in common with other biennials, does not withstand cultivation.

Similar plants   Although all thistles have spines on the leaf edges, bull thistle is the only thistle with a spiny leaf surface. Bull thistle has winged stems with long pointed prickles; this character is sufficient for differentiation from all other true thistles, *Cirsium* spp., except marsh thistle, *Cirsium palustre* (L.) Scop., a European weed known in Canada only in Newfoundland, Nova Scotia, and at Prince Rupert, British Columbia. The plumeless thistles, *Carduus* spp., also have spiny-winged stems, but differ from the true thistles in having a pappus of simple hairs.

There are a number of native thistles. Two of these, wavy-leaved thistle, *Cirsium undulatum* (Nutt.) Spreng., and Flodman's thistle, *Cirsium flodmanii* (Rydb.) Arthur, are mistaken for the bull thistle, although they are clearly different from that species in having densely woolly and wingless stems. Wavy-leaved thistle is a plant of southern British Columbia, Alberta, and Saskatchewan. Flodman's thistle ranges across the three Prairie Provinces from the Canadian border to the northern limit of settlement. It apparently does not occur in British Columbia.

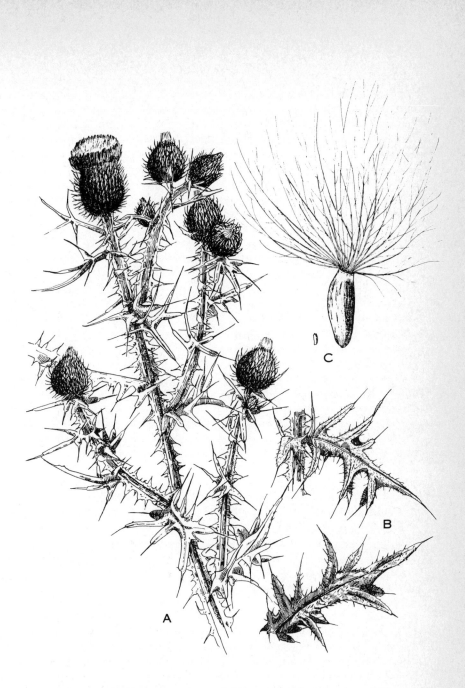

Bull thistle, **Cirsium vulgare: A,** plant; **B,** leaf showing the spiny surface; **C,** seed and pappus.

COMPOSITE FAMILY — COMPOSITAE

RUSSIAN KNAPWEED *Centaurea repens* L.

Other scientific names   *Centaurea picris* Pall. and *Acroptilon picris* (Pall.) DC.

Description   Perennial, forming dense patches, spreading by seeds and creeping horizontal roots; roots blackish and scaly, producing stem buds that develop into shoots; stems branching from near the base, erect, usually 6-9 dm high, young stems covered with soft gray hairs or nap; leaves alternate, hairy at early stages, more or less clasping, lower leaves elongated and notched, upper leaves progressively smaller and more entire towards the top of the plant; flower heads almost spherical, characteristically silvery when unopened, about 12 mm in diameter, solitary at the ends of rather long leafy branches; bracts of the head entire, greenish at the base, and with a papery, finely hairy tip; flowers numerous, all tubular, purplish or pink in bloom, and straw-colored at maturity; seeds 3 mm long, compressed, glabrous, chalky white, flattened, oval in outline, and with a basal scar; pappus of stiff, whitish, rough bristles, the inner bristles are about 9 mm long and double the length of the others, bristles soon deciduous. Flowering in July.

Origin   Southern Russia and Asia Minor to Altai and Afghanistan. Introduced with imported Turkestan alfalfa.

Distribution in Canada   A recent introduction, now widely distributed in southwestern Manitoba, Saskatchewan, and Alberta, and quite common in south central British Columbia. Several infestations are known in southern Ontario.

Habitats   Cultivated fields, grain and alfalfa fields, pastures, and waste places.

Notes   There is evidence that Russian knapweed is poisonous to sheep and horses. A South African worker reported that a full-grown sheep died 18 hours after it was fed just over half a pound of dried plant. A Russian text claims that horses have been poisoned. No losses from this plant have been reported in Canada.

Similar plants   Russian knapweed resembles the thistles, but has smaller heads and is without prickles.

The flower heads of Russian knapweed have entire bracts, while the bracts in the other species of knapweed have fringed or torn tips. The seeds in Russian knapweed have a scar at the base, and the pappus is longer than the seed. In the other species, the seed has a notch at one side and the pappus is much shorter than the seed. The blackish scaly roots of Russian knapweed are also distinctive.

190

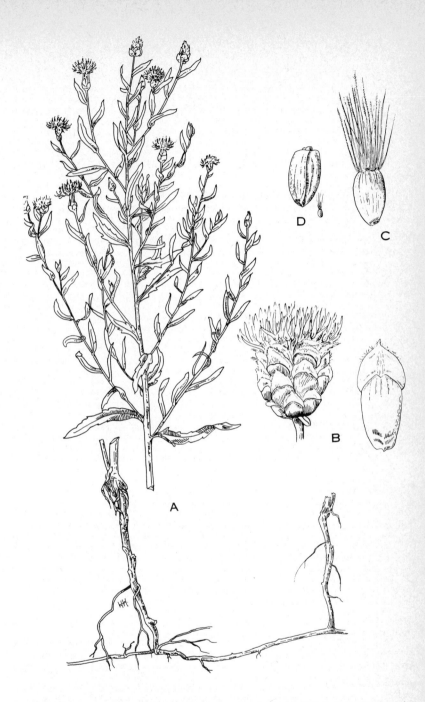

Russian knapweed, **Centaurea repens: A,** plant; **B,** head of flowers and to the right a bract of the head; **C,** seed with pappus; **D,** seed with pappus removed.

COMPOSITE FAMILY — COMPOSITAE

DIFFUSE KNAPWEED *Centaurea diffusa* Lam.

Description   Biennial to short-lived perennial; stems 6-9 dm high, branched; leaves alternate, much divided, rough hairy; flower heads numerous, small, and narrow; bracts of the head ending in a rigid, often spreading spine; flowers white or purplish; seeds about 2.5 mm long; pappus lacking or a mere fringe about 1/6 the length of the seed.

Notes   This Eurasiatic plant is only common in Canada in southern British Columbia.

Easily mistaken for spotted knapweed. In diffuse knapweed the bracts end in a rigid spreading spine. In spotted knapweed the bracts lack these long spines and always have a terminal blackish fringe.

SPOTTED KNAPWEED *Centaurea maculosa* Lam.

Description   Biennial or short-lived perennial; stems 6-9 dm high; branched; leaves alternate, deeply cut into narrow divisions, somewhat hairy; heads numerous; bracts of the head with a black-fringed tip; flowers purple, rarely white; seeds 3 mm long; pappus of simple bristles 1/4 to 1/2 the length of the seed.

Notes   This European weed is found in Nova Scotia, New Brunswick, Quebec, Ontario, and British Columbia. It is probably most abundant in British Columbia, but there are large stands in Ontario, particularly in Grey and Hastings counties.

BLACK KNAPWEED *Centaurea nigra* L.

Description   Perennial; stems erect, 6-12 dm high, branched; leaves alternate, hairy, undivided, irregularly toothed, the lower leaves with stalks, the upper leaves narrow and sessile; heads about 25 mm across, rounded and hard; bracts of the head straw-colored at the base, blackish or brownish fringed above; flowers purple; seeds sparsely hairy, oblong, compressed; pappus very short.

Notes This European plant is abundant in the Atlantic Provinces. It is also known from more inland points in Quebec and Ontario and in British Columbia.

BROWN KNAPWEED *Centaurea jacea* L.

Description   Perennial; stems erect 6-12 dm high, branching near the top; leaves alternate, undivided, irregularly toothed, rough hairy, the basal leaves with stalks, the upper leaves smaller and sessile; bracts of the head pale brown, the broad upper part appearing to be torn into irregular divisions; flowers rose-purple, the marginal row usually larger; seeds 3 mm long, sparsely hairy, compressed; pappus lacking.

Notes   This European plant is known from Ontario and Quebec. It is particularly common in southwestern Ontario and the Eastern Townships of Quebec.

Brown knapweed and black knapweed have undivided leaves and therefore are easily distinguished from the other two species mentioned.

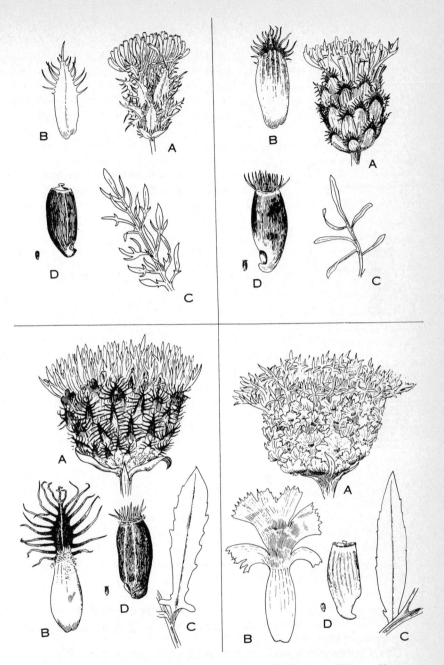

Knapweeds. **Upper left:** Diffuse knapweed, **Centaurea diffusa. Upper right:** Spotted knapweed, **Centaurea maculosa. Lower left:** Black knapweed, **Centaurea nigra. Lower right:** Brown knapweed, **Centaurea jacea.** In all drawings: **A,** head of flowers; **B,** a single bract of the head; **C,** leaf; **D,** seed.

COMPOSITE FAMILY — COMPOSITAE

CHICORY *Cichorium intybus* L.

Other names   Blue daisy, blue sailors, coffee-weed, common chicory, wild succory.

Description   Perennial with milky juice, spreading by seeds; taproot long and fleshy; stems erect, hollow, 3-15 dm high, with stiff spreading branches, usually rough hairy; leaves alternate, rough hairy, those of the base and lower stem deeply cut, the upper leaves far smaller, clasping the stem, gradually becoming reduced and bractlike in the inflorescence; heads 4 cm broad in flower, stalkless or at the tips of the short branches, usually with long-stalked glands; outer bracts of the head white and thickened in the lower half, green above; flowers several in each head, large, bright blue, rarely pink or white, all ligulate; seeds 2.5 mm long, irregularly angular, grooved and ridged lengthwise, straw-colored to brown, and often with darker mottling; pappus white, scaly, and very short, scarcely 1/10 as long as the seed, forming an insignificant crown. Flowering from July to frost, the heads usually close by midday.

Origin   Europe.

Distribution in Canada   Reported from every province. Abundant in Eastern Canada and often found in southern British Columbia. From Sault Ste. Marie to the Alberta – British Columbia border it is an extremely rare plant: there are less than 10 records for this huge area. Chicory is far commoner in the more southerly agricultural areas and is of little consequence in northern settlements or in acid soils such as those of the Eastern Townships of Quebec.

Habitats   Hayfields, roadsides, waste places, and fencerows. Usually on high-lime soils.

Notes   The roots of this plant are dried, roasted, and ground and used commercially as a substitute for coffee, or as an adulterant. A small percentage of chicory in coffee is preferred by many and there is evidence that chicory helps retain the aromatic constituents of coffee.

If eaten in large quantities by cows, the herbage imparts a bitter flavor to milk.

Similar plants   Chicory in flower is not likely to be mistaken for any of its relatives except possibly blue lettuce (page 204). The basal leaves form a rosette that superficially resembles that of the dandelion.

Chicory, **Cichorium intybus: A,** plant; **B,** dark- and light-colored seeds.

FALL HAWKBIT *Leontodon autumnalis* L.

Other names   August flower, fall dandelion. Also known scientifically as *Apargia autumnale* (L.) Hoffm.

Description   Perennial with milky juice, spreading by seeds; rootstock short and thick, sometimes branched, each branch ending in a rosette; stems 10-60 cm high, branched above; leaves all at the base except for scaly bracts on the stem, basal leaves usually rather deeply divided, glabrous or with a few simple hairs; flowering heads several on each stem, borne singly at the ends of branches, 18-25 mm across; bracts of the head dark green, in 2 rows, one main inner row of long narrow bracts and a much shorter outer row, glabrous or somewhat pubescent, in var. *pratensis* (Link) W.D.J. Koch the bracts covered with blackish hairs, flowers all of one kind, ligulate, golden yellow; seeds narrowed to the top but not beaked, 3-6 mm long, brownish, ridged lengthwise and with numerous finer ridges across; pappus a single row of feathery hairs, tawny. Flowering from early July to late September.

Origin   Europe, Asia, and northwest Africa.

Distribution in Canada   Common in the Atlantic Provinces and found in Ontario and Quebec. Rapidly spreading in Quebec, particularly in the Eastern Townships.

Habitats   Old pastures, meadows, roadsides, lawns, and waste places. Fall hawkbit appears to be particularly well adapted to the invasion of well-grazed permanent pastures.

Similar plants   Both fall hawkbit and the dandelion, *Taraxacum officinale* Weber, have basal rosettes of deeply cut leaves, and both plants have large yellow flowers. Fall hawkbit has a scaly bracted and branched stem, while, in the dandelion, no stem is present and the hollow, naked flower stalks bearing single heads rise directly from the basal rosette. The illustration on the opposite page shows that the seeds are very different in shape and that the seed in dandelion has a long beak supporting a pappus of simple hairs, whereas the seed in fall hawkbit is beakless and the pappus is feathery.

Spotted cat's ear, *Hypochaeris radicata* L., a common plant in the lower Fraser Valley of British Columbia and known from Eastern Canada, resembles fall hawkbit in having a basal rosette and scaly bracted stems. Spotted cat's-ear has coarser, hairier, less divided leaves and strongly beaked seeds.

Fall hawkbit, **Leontodon autumnalis: A,** plant; **B,** head of flowers; **C,** mature head; **D,** seed and pappus. Dandelion, **Taraxacum officinale: E,** plant; **F,** seed showing beak and pappus and enlarged seed without beak or pappus.

MEADOW GOAT'S-BEARD *Tragopogon pratensis* L.

Other names Johnny-go-to-bed-at-noon, meadow salsify, yellow goat's-beard.

Description Biennial and perhaps perennial with milky juice; taproot long and fleshy; stems simple or branched, 3-9 dm high; leaves alternate, grass-like, margins crisped, base clasping and enlarged, stem leaves contracting abruptly into long tips, which are curled backwards; heads large, solitary, and terminal on stalks that are only slightly enlarged at the summit; bracts of the head about 8 in number, sometimes with reddish margins, at flowering time less than 25 mm long, elongating slightly as seeds mature; all flowers ligulate, yellow, usually longer than the bracts, 5-notched at the tip; seeds tapering into a slender beak, seeds including the beak less than 25 mm long, ribbed, the inner seeds of the head smooth, the outer seeds coarsely roughened; pappus of numerous bristles, which have hairlike side branches (plumose), whitish or tawny. Flowering from late May to September but mainly in June and July, closing by noon on bright days.

Origin All the goat's-beards are from Europe.

Distribution in Canada The yellow-flowered goat's-beards occur in every province except Newfoundland. *Tragopogon pratensis* is rare in the Prairie Provinces and common in Eastern Canada. Western goat's-beard, *Tragopogon dubius,* is abundant in the Prairie Provinces and is now spreading rapidly in Ontario and is established in Quebec.

Habitats Old meadows, roadsides, along railways, and waste places.

Similar plants Western goat's-beard, *Tragopogon dubius* Scop. (*Tragopogon major* Jacq. in some floras), has been mistaken for *Tragopogon pratensis.* The former plant may be distinguished from the species illustrated and described by the following characters: that stalk below the flower head is very thick, hollow, and inflated; the leaves are not curled at the apex nor crisped at the margins; the flowers are almost always shorter than the bracts; the seeds including the beak are about 30 mm long; the bracts are more numerous and longer, particularly at maturity, when they average about 50 mm long.

Common salsify or oyster-plant, *Tragopogon porrifolius* L., has purple flowers but otherwise resembles the other goat's-beards. This plant sometimes escapes from gardens, where it is grown as a vegetable.

Meadow goat's-beard, **Tragopogon pratensis: A,** plant; **B,** mature head; **C,** seed with pappus; **D,** seed natural size and without pappus.

PERENNIAL SOW-THISTLE *Sonchus arvensis* L.

Other names   Creeping sow-thistle, field sow-thistle.

Description   Vigorous, deep-rooted perennial with milky juice, spreading by seeds and by fleshy, horizontal, easily broken roots that bear new buds at many intervals; stems erect, 6-15 dm high, hollow, glandular hairy or glabrous above; leaves variable in outline, margins with small weak prickles; basal leaves narrowing to a winged stalk, deeply cut into lobes, which often curve backwards; stem leaves alternate, similar to the basal leaves but sessile and clasping the stem with rounded basal lobes; flower heads up to 4 cm across when expanded, in a loose terminal inflorescence, heads and branches of inflorescence densely covered with stalked glandular hairs or glabrous (var. *glabrescens* Guenth., Grab. & Wimm.); bracts dark green or lead-colored; flowers bright yellow, numerous in each head, all ligulate; seeds up to 3 mm long, dark brown, ribbed, and cross-wrinkled; pappus white, of numerous simple hairs. Flowering from June to September.

Origin   Europe, Caucasus, Asia Minor, Afghanistan, West Siberia. Widely introduced in Asia, Africa, Australia, and America.

Distribution in Canada   A common weed in all provinces. Particularly abundant in Ontario, Manitoba, and the more northerly agricultural districts of Quebec, Ontario, and the Prairie Provinces.

Habitats   Cultivated fields, grainfields, waste ground, roadsides, river and lake shores. Grows under a wide variety of soil conditions but thrives best in moist fertile soils.

Notes   This plant spreads so rapidly that it occurred in 1950 on the road from The Pas to Flin Flon, which was completed in 1949. Indicative of its recent introduction was the fact that it did not form large patches, but occurred as a few plants every few hundred yards.

Similar plants   Perennial sow-thistle may be easily distinguished from the annual sow-thistles by its creeping roots, larger flower heads, and, in the glandular variety, the abundant sticky hairs on the bracts of the flower heads.

Blue lettuce (page 204) is similar to sow-thistle, but has blue flowers and very different seeds. The leaf edges are without the soft weak prickles of perennial sow-thistle.

Prickly lettuce (page 206) is yellow-flowered, but the flower heads are much smaller and the midribs have sharp bristles.

Perennial sow-thistle, **Sonchus arvensis: A,** plant; **B,** seed.

SPINY ANNUAL SOW-THISTLE *Sonchus asper* (L.) Hill

Other names   Prickly annual sow-thistle, spiny-leaved sow-thistle.

Description   A taprooted annual with milky juice, spreading by seeds; stems 3-12 dm high, very often with a few scattered, stalked, glandular hairs in the upper part; leaves usually dark glossy green above, leaf margins usually crisped and very sharply and prominently toothed, spiny to the touch, leaves rarely divided; upper stem leaves sessile, clasping the stem with markedly ear-shaped and rounded lobes; flower heads flask-shaped, less than 2.5 cm across; flowers all ligulate, light yellow; seeds barely 3 mm long, brown, flattened, with a few prominent parallel ribs on each surface, without cross-wrinkles; pappus white, of numerous simple hairs. Flowering June to September.

Origin   Europe.

Distribution in Canada   Known from all provinces. More abundant in Ontario, Quebec, and British Columbia than in the other provinces.

Habitats   Cultivated fields, gardens, roadsides, and waste places.

Similar plants   This species differs from the annual sow-thistle in the following ways: leaves rarely divided; leaf margins very prickly; ears at the base of the upper leaves rounded not pointed; seeds without cross-wrinkles.

ANNUAL SOW-THISTLE *Sonchus oleraceus* L.

Other name   Common sow-thistle.

Description   A taprooted annual with milky juice, spreading by seeds; stems 3-12 dm high, often with a few scattered stalked glandular hairs in the upper part; leaves usually divided into lobes, the terminal lobe large and triangular; leaf margins with small weak teeth; basal leaves stalked; stem leaves sessile and clasping the stem with pointed lobes, all stem leaves deeply divided except sometimes the uppermost; flower heads less than 25 mm broad, white downy at the base when in bud; flowers all ligulate, light yellow; seeds barely 3 mm long, reddish brown, ribbed, and cross-wrinkled; pappus white, of simple hairs. Flowering June to early October.

Origin   Europe.

Habitats   Cultivated fields, gardens, roadsides, and waste places.

Distribution in Canada   The same as for spiny annual sow-thistle.

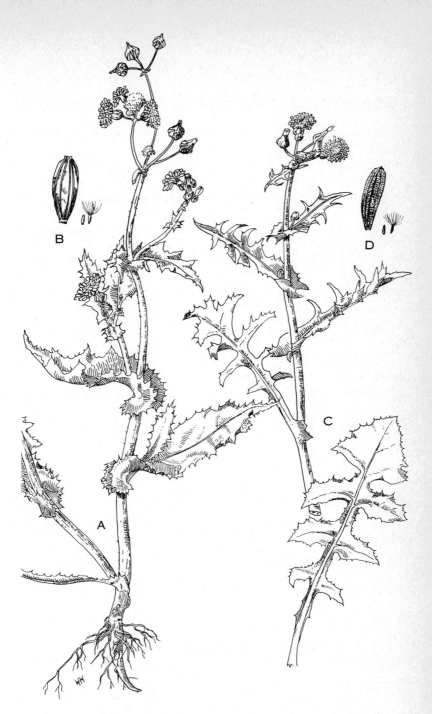

Spiny annual sow-thistle, **Sonchus asper: A,** plant; **B,** seed. Annual sow-thistle, **Sonchus oleraceus: C,** plant; **D,** seed.

COMPOSITE FAMILY — COMPOSITAE

BLUE LETTUCE *Lactuca pulchella* (Pursh) DC.

Other names   Blue flowering lettuce, perennial lettuce, showy lettuce, wild blue lettuce.

Description   Perennial, with milky juice, spreading by seeds and rather deep rootstocks that send up numerous shoots; stems rarely over 9 dm high, somewhat branched above; leaves alternate, sessile, the lower leaves 5-15 cm long, usually coarsely toothed with the divisions directed backward, the upper leaves smaller and entire; heads large, over 25 mm across in flower, bracts of the head of various lengths, the inner to 18 mm long; flowers blue, showy, all ligulate; seeds flattish, up to 6 mm long including the beak, strongly nerved on the flat surface, tapering to a short, definite beak, red when immature, slaty-gray when ripe; pappus copious, white, longer than the seed. Flowering from June to August, flowers remaining open until late afternoon.

Origin   Native to western North America. Introduced into Europe (Sweden).

Distribution in Canada   Practically confined to the Prairie Provinces and British Columbia. It ranges northwestward across the prairies to the Peace River district and has been collected far to the north on Great Bear Lake in the District of Mackenzie. This plant has spread from native habitats to cultivated land in the west, but is not encroaching rapidly eastward as evidenced by the very few records for Ontario and a single report for Quebec.

Habitats   Open prairie, river shores, alkaline flats, meadows, roadsides, railroad rights of way, waste places, and irrigated fields.

Note   Blue lettuce is an excellent example of a weedy native plant that has increased in abundance with agricultural expansion.

Similar plants   Tall blue lettuce, *Lactuca biennis* (Moench) Fern., is an annual or biennial plant reaching 3 m high. It also differs from blue lettuce in having all leaves coarsely toothed, more numerous and smaller heads, and a brownish pappus.

Chicory is sometimes mistaken for blue lettuce and the illustration and description of chicory (page 194) should be referred to if there is any doubt. Chicory is largely eastern in distribution and blue lettuce almost entirely western.

Blue lettuce, **Lactuca pulchella: A,** plant; **B,** seed.

COMPOSITE FAMILY — COMPOSITAE

PRICKLY LETTUCE *Lactuca scariola* L.

Other names Compass plant, wild lettuce. Another scientific name used is *Lactuca serriola* L.

Description Annual or winter annual with milky juice, taprooted; stems 3-15 dm high, often with sharp prickles below, or smooth; stem leaves alternate, mostly deeply divided, or all leaves undivided in forma *integrifolia* (Bogenh.) Beck, stalkless, clasping the stem by pointed earlike lobes, leaf margins spiny-toothed, the whitish midvein on the lower surface of the leaf bearing a row of stiff, sharp prickles; flower heads small, numerous, in a leafy open inflorescence; bracts of the head narrow and glabrous, elongating at maturity; flowers all ligulate, few in each head, pale yellow (drying bluish); seeds enlarged upwards, contracting abruptly to a threadlike, often twisted, beak that is somewhat longer than the body of the seed, seed and beak 6-8 mm long; body of the seed flattened, 5-7 ribbed, often mottled, roughened, and with a few white bristles above; pappus of simple hairs about 4 mm long. Flowering from mid-July to mid-September.

Origin Europe.

Distribution in Canada Occurs in every province except Newfoundland. Very rare in Prince Edward Island, Nova Scotia, and New Brunswick. Abundant in southern Ontario and common in the Prairie Provinces.

Habitats Cultivated land, roadsides, waste places, and gardens.

Notes Prickly lettuce is closely related to garden lettuce. Usually the leaves of prickly lettuce are twisted so that the leaf blade is vertical rather than horizontal. The broad surface of the leaf faces the rising and setting sun so that at noon the leaf is edged to the sun. This arrangement must protect the plant from the most intense heat yet permit full insolation when the sun is towards the horizon. The tips of the leaves point north or south and this gives rise to the name "compass plant."

Similar plants The spiny leaf margins and midveins should be sufficient to distinguish this plant from others that resemble it.

Prickly lettuce, **Lactuca scariola: A,** plant; **B,** seed with and without pappus and beak. Entire-leaved prickly lettuce, **Lactuca scariola** forma **integrifolia: C,** leaves.

COMPOSITE FAMILY — COMPOSITAE

NARROW-LEAVED HAWK'S-BEARD *Crepis tectorum* L.

Description   Annual or winter annual with milky white juice; flowering stems 7 cm to 9 dm high; leafy; basal and lower leaves very variable, usually sparingly toothed or lobed, upper stem leaves narrow and entire, sessile, sometimes clutching the stem by small pointed lobes; flower heads up to 18 mm across when full expanded, much smaller before and after flowering; flowers all ligulate, bright yellow, longer than the bracts; bracts in several rows with the outer row much shorter than the inner, inner bracts with short appressed hairs on the inner surface; seeds dark purplish when mature, 3 mm long, strongly ribbed; pappus of simple hairs, soft and white. Flowering from mid-June throughout the summer.

Origin   Eurasia.

Distribution in Canada   Widely distributed in Canada, occurring in every province except Newfoundland. Abundant in southern Manitoba from the United States border to Dauphin, and found in the Interlake region. There is evidence that it is spreading in Saskatchewan, Alberta, and British Columbia. The extent of the infestations in Manitoba suggests that this plant may become a much more serious menace in other parts of Canada.

Habitats   Grainfields, pastures, fallow land, roadsides, and railway yards.

Similar plants   Smooth hawk's-beard, *Crepis capillaris* (L.) Wallr., of European origin, is a rather rare plant known only from a few locations in Nova Scotia, Quebec, Ontario, and British Columbia. In British Columbia it appears to be confined to Vancouver Island and the extreme southwestern corner of the mainland. Smooth hawk's-beard might be mistaken for narrow-leaved hawk's-beard, but the bracts of the head are not hairy on the inner surface, the seeds are pale brown (not purple), the heads are smaller, and the stem leaves clutch the stem by prominent pointed lobes.

The weedy hawk's-beards resemble the hawkweeds, but are separable by being annual or winter annual in habit and in having a whitish pappus. The hawkweeds are fibrous-rooted perennial plants and the pappus is brownish.

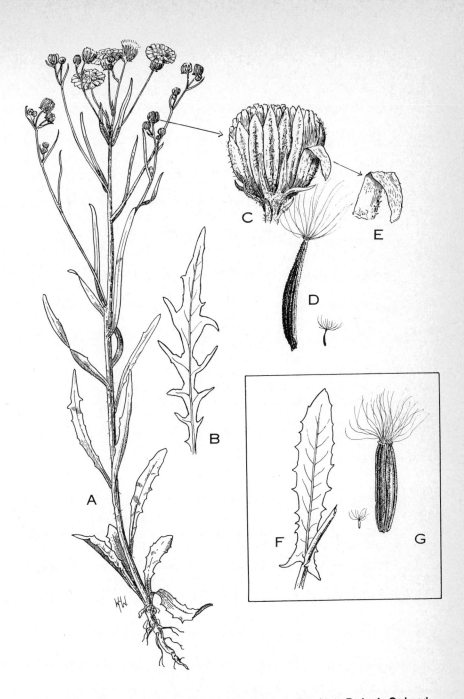

Narrow-leaved hawk's-beard, **Crepis tectorum: A,** plant; **B,** leaf; **C,** head of flowers; **D,** bract of head turned back to show fine hairs on inner surface; **E,** seed. Smooth hawk's-beard, **Crepis capillaris: F,** leaf; **G,** seed.

ORANGE HAWKWEED *Hieracium aurantiacum* L.

Other names   Devil's paint-brush, orange paint-brush.

Description   Perennial with milky juice, spreading by seeds and leafy runners; flowering stems 20 cm to 6 dm high, hairy, usually leafless, sometimes with 1 or 2 reduced leaves; leaves largely in a basal rosette, entire, hairy; flower heads nearly 25 mm across, in a crowded cluster at the top of the stem; flowers all ligulate, bright orange-red; bracts of the head in 2 or 3 rows, blackish hairy; seeds about 2 mm long, deep red when immature, purplish black when mature, ridged lengthwise; pappus about 3 mm long, dirty white, of simple hairs. Flowering from June to August.

Origin   Europe.

Distribution in Canada   Very abundant in Ontario and Quebec and apparently spreading in the Atlantic Provinces. Rare in British Columbia and only known from single collections in Alberta and Manitoba.

Habitats   Old fields and pastures, meadows, roadsides, and open woods. Usually on poorer acid soils.

Notes   Since most of the hawkweeds produce seed without fertilization of the ovary, cross-breeding is prevented and slight individual differences are perpetuated. Some European botanists have specialized in recognizing these differences, and therefore many thousands of "species" and other groups have been named. Identification of the hawkweeds introduced to this continent is thus difficult, and must, as yet, be arbitrary. At present, the best approach is to follow one of the more recent floras such as the Eighth Edition of Gray's Manual of Botany.

Similar plants   A number of weedy hawkweeds of European origin occur in Canada. Only orange hawkweed has orange-red flowers. All the other species have yellow flowers.

King devil hawkweed, *Hieracium florentinum* All., has very short, thick rootstocks, stems without leaves, or with 1 or 2 reduced leaves, lacks aboveground runners, and the leaves are only sparingly hairy (see illustration). King devil hawkweed is abundant in Ontario and is spreading rapidly into western Quebec.

Mouse-ear hawkweed, *Hieracium pilosella* L., differs from all the other hawkweeds in having only 1 flower head on each stem. This plant has aboveground runners and forms dense mats. It is common in the Atlantic Provinces and is spreading in Ontario and Quebec.

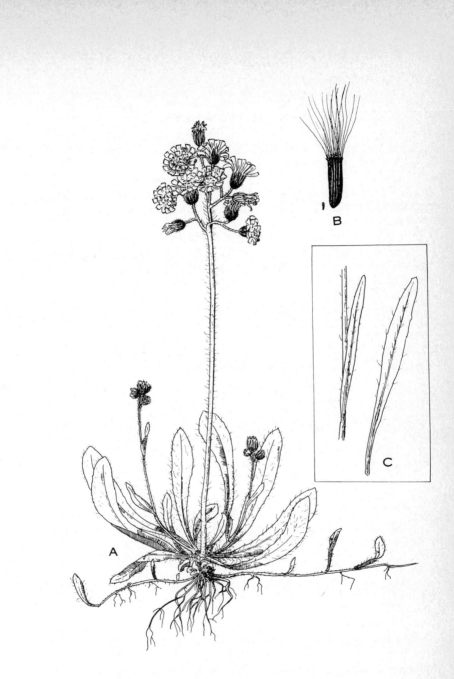

Orange hawkweed, **Hieracium aurantiacum: A,** plant; **B,** seed. King devil
hawkweed, **Hieracium florentinum: C,** leaves.

The scientific names and the pages with illustrations are in boldface.